AF615840

Solar Power
Energy of the Future

Edited by

Nirmala Rao Khadpekar

2012

Icfai Books
The Icfai University Press

SOLAR POWER: ENERGY OF THE FUTURE

Editor: Nirmala Rao Khadpekar

© 2012 The Icfai University Press. All rights reserved.

No part of this publication may be reproduced, stored in a retrieval system, or transmitted in any form or by any means - electronic, mechanical, photocopying or other otherwise - without prior permission in writing from the Icfai University Press.

While every care has been taken to avoid errors and omissions, this publication is being sold on the condition and understanding that the information given in the book is merely for reference and must not be taken as having authority of or being binding in any way on the authors, editors, publishers or sellers.

Icfai Books, IB and IB logo are trademarks of the Icfai University Press. Any other product or corporate names, that may be registered trademarks, are used in the book only for the purpose of identification and explanation, without any intent to infringe.

First Edition: 2012
Printed in India

Published by

This book is published by IUP.
University Campus, Agartala-Simna Road,
P.O. Kamalghat Sadar, Agartala – 799210, Tripura (West)
E-mail: info@iupindia.org
Website: www.books.iupindia.org

Unless repugnant to the context otherwise, any reference to the words Icfai, Icfai Books, Icfai University Press shall be read and construed as IUP only. This book is not for sale in US and Canada.

ISBN: 978-81-314-2739-2

The views and content of this book are solely of the author(s)/editor(s). The author(s)/editor(s) of the book has/have taken all reasonable care to ensure that the contents of the book do not violate any existing copyright or other intellectual property rights of any person in any manner whatsoever. In the event the author(s)/ editor(s) has/have been unable to track any source and if any copyright has been inadvertently infringed, please notify the publisher in writing for corrective action.

CONTENTS

Overview

The sun is credited with being a free, abundant and inexhaustible source of solar energy which has the advantage of having a low environmental footprint. Conventional power generating and transmission technology is becoming increasingly expensive to manage in terms of high investment in new reserves while the demand for power is always rising. While the solar energy capital and generation cost is currently high, three to four times more expensive than conventional energy, the costs in this sector have been declining on an average of four per cent per annum, considered one of the fastest declines of any energy source. This decline in costs is driven by continuous advances in photovoltaic technology and manufacturing economies of scale. While global renewable energy capacities grew at the rate of 15-20 per cent annually for many technologies, grid connected solar PV technology recorded a 60 per cent annual average growth for the five-year period beginning 2002. Worldwide, solar energy industry is doubling in size every 20 months and will reach $100 billion per year by 2011.

Energy utilities and governments in most countries, including India, are putting emphasis on solar among the renewables. India's National Action Plan on Climate Change released on June 30, 2008, focuses on eight priority national missions, the first among which is solar energy, whose potential, according to the Prime Minister, will change the face of the country. This reaffirms India's pledge to pursue sustainable development along with a cap on GHG emissions despite the country's development imperatives.

Solar energy, like wind, is an attractive option from among the renewable resources, as post initial investment in equipment, it has low maintenance and close to zero variable costs, and is not exposed to price volatility. Technology for harnessing solar energy which has been active over the last ten years has been receiving increased attention due to the impact of the energy crisis on the economy. The market, however, for solar energy alternatives is still evolving, needing intelligent choices to be made by venture capitalists, utilities and government regulatory bodies to enable nurturing of this opportunity. Some countries like Spain are bringing in solar guidelines into standard building norms to meet the energy challenge. Markets like California and Italy have achieved economic competitiveness in a relatively short period and their examples could be emulated to profit.

Of late, though there has been a renewed interest in energy, there is a tendency to think about renewables like solar energy as a future possibility to be looked at seriously only when traditional sources run out. With the oil prices hitting lows at the time of writing, there is a probability that the renewed interest will again rest a while. To keep up the pressure to make a real change in the energy scenario this book is presented in three sections – Value Proposition, Technology Initiatives, and Experiences and initiatives, to take forward the argument.

The first section titled **Value Proposition** has five articles.

The article titled **"Solar: The Most Important Renewable of the Future"**, by *Andrew Blakers* explores the idea of how energy efficiency, renewable energy, carbon capture and storage ('clean coal'), afforestation and other measures need to be harnessed to solve the fallouts of climate change. None of these options will prosper without a price on carbon pollution. The author works out why this shift is important as renewable energy and energy efficiency are actually more job intensive than coal fired power stations. He says since renewable energy comprises many energy forms including to a large part solar energy generation and clean transport systems, taken together renewable energy and energy efficiency can provide most human needs by 2050. Direct solar energy such as photovoltaic and solar thermal, with a resource 1,000 larger than current energy consumption, is the most important renewable energy in the long term. Based in Australia, the author uses the country as spring point to make his worldwide observations. It is important to have a balanced portfolio with increased support for renewables with targets and energy efficiency R&D, demonstration, commercialization and market incentives with carbon pricing and individual technology incentives. The illustration of Australia is given in this article.

The second article in this section is titled "**Payback and other Financial Tests for Solar Electric Systems**" by *Andy Black*. As the financial case is a major deciding factor for most people putting out a fresh investment; the payback question can be given an answer by examining high electricity rates, financial incentives, net metering policies and good sunlight, says the author. The answers vary by local climate, utility rates in places like California; nevertheless yielding a 10 per cent plus annual rate of return. Property resale values also cover PV system investment in most places but there are places, where the conditions barely cover maintenance costs. Then net metering policies are discussed which operate on time-of-use-rate-schedule which end up varying the price by time of day; with solar producing high during day time load requirement, the payback factor improves. Performance based incentives and renewable energy credit combinations can improve the economics further. Inflation in electricity rates does not affect as in case of solar which is considered an inflation protected investment. It offsets electricity costs at the prevailing electricity rates letting savings actually grow and improve cash flows. The key to proportion of solar technologies and application is to create markets that reward investment. Certain markets are already exploding as PV system owners have discovered the financial benefits.

The third article is a case study titled "**Exploring the Economic Value of EPAct 2005's PV Tax Credits**" by *Mark Bolinger, Ryan Wiser* and *Edwin Ing*. This 2007 update is from the Berkeley Clean Energy States Alliance which tracks the impact of the Energy Policy Act of 2005 which supports both commercial and residential PV systems to drive even faster growth over the next decade. The report examines the value and explores the implications using a generic cash flow model and the interaction between governmental credits and grants for different types of entities, like purchasers and administrators. The findings also have important implications for policy design, and the type of incentives offered depending on capacity and performance, larger residential systems, and utility systems, as it will have an effect on the nature of consumer demand for PV with credits. Preference of commercial system owners for taxable grants and residential system owners for non-taxable grants are discussed with possibilities of third party ownership evolving to capture the benefits of new credits.

The fourth article is titled "**Brokering for Reducing Costs and Increasing Customer Satisfaction in Solar Sales**" by *Andy Black*. The service brokering model for sale of solar systems is examined here in its similarity and dissimilarity to other brokering, benefits and risks, savings and satisfaction levels. Sales people offer limited options with limited knowledge; while a trained, knowledgeable and experienced broker could do a sterling service, feels the author who has developed a unique and successful business model. The author has developed a series of screening techniques involving the bid specs and normalizes them with less salesmanship and more solid stuff to enable customer decision-making. The author says that it is beneficial for both the installers and purchasers to go through a known and reliable agent for time saving and effort reduction as aggression on price is not fruitful as the solar industry does not enjoy very high profit margins. Risks in brokering and risk mitigation to the purchasers and installer are discussed, which the broker is expected to minimize. A cautionary note is added about consumers needing to check broker referrals from people who have lived with their systems for at least a year. Gains in efficiency are expected all around considering that solar industry by estimate reduces costs at 4.5 per cent a year (based on 35 per cent compound annual growth in production volume that has occurred annually over the last seven years). He feels that the broker concept could improve those percentages with successful and pertinent installations and jump ahead the experience cost curve further appreciably.

The fifth and last article in this section titled **"Solar Energy: Gearing up to Cope"** by *Nirmala Rao Khadpekar* discusses how the solar energy perceived uneconomic is gaining acceptance worldwide as renewable technologies improve with economies of scale from cookers to satellite technology. The economics of the sector is changing though solar energy capital and generation cost is currently high, three to four times more expensive than conventional energy. Costs in this sector have been declining on an average of 4 per cent per annum. Subsidies have played a role in all forms of power supply, which makes comparisons odious sometimes but there is continued need for governmental support and direction as true grid parity is not yet in sight. Developing countries have a different set of priorities related to economics before environment. Solar power is already creeping towards competitiveness in certain areas. Clarity of objectives, reward for production, not capacity with gradual subsidy fade out is needed. Competition in evolving technology is attracting venture capital and strategic alliances. The value proposition of solar power is discussed, along with the social and environmental costs. Climate change and unexpected variants are making investment in energy security imperative as a societal matter. IPPC is encouraging all countries to redress this. A realistic carbon tax would adjust the real cost and viability of renewables like solar.

The second section titled **Technology Alternatives** has six articles.

The article on **"Technology Advances in Delivering Cost-Competitive Solar Energy"** comes from the legendary R&D address—the Palo Alto Research Centre. It describes a particular high performance CPV solution which compared to the average flat PV plate, uses 1/500th as much

PV material and produces twice as much electricity for a given collection area. The paper outlines how the silicon-based flat plate Photovoltaic using semiconductor PV material converts sunlight directly into electricity for generating solar electricity and defined the solar electricity market since its emergence 30 years ago. However, recent silicon feedstock shortages and rising wafer prices have highlighted the need to advance the market beyond current flat-plate PV and its underlying cost limitations. One option is concentrator photovoltaic (CPV) that uses mirrors or lenses to focus sunlight collected from a given receiver area onto a smaller area of PV material, and in some cases uses higher efficiency, non-silicon PV material. The proposed solution is also smaller, cheaper and easier to manufacture. These improvements help generate electricity at less than half the cost possible with existing flat-plate technologies, potentially opening large new markets for clean solar energy.

The next article is titled **"2008 Photovoltaic Trends: Innovative Thin Film Technology and Large-Scale Power Plants; Gigawatt Perspectives in the USA"** by *Rolf Hug*. It discusses large scale solar plants for megawatt electricity generation and the new thin film PV technologies and products which are proliferating the world with more and larger PV plants coming up in Europe, America and Asia. In particular, turnkey solutions for manufacture of thin film modules, solar cells and solar module manufacturing systems are growing in the global market. This report from Solarserver, which is a Forum for Solar Energy, is official media partner for conferences for Intersolar 2008, and Solar Gigawatts USA, focuses on current trends in the thin film PV market and spotlights solar markets in North America. The report notes how companies from the electronics industry (semiconductors and plasma technology) are increasingly focusing on PVs. Swiss Oerlikon and Japanese Sharp, Sanyo and Kaneka have developed turnkey systems for thin film modules to take over the extensive planning processes and the establishment of the manufacturing lines for future producers. The next logical step is for industrial manufacture of thin film technology to enhance the competitiveness of solar power generation and further integrated production systems which are the international trend. A run through of various worldwide companies reveals that 80 per cent of all large scale commercial PV are located in Europe (Germany 50 per cent of world) with positions 2 and 3 being held by the USA and Asia.

The following article is titled "**Solar Energy Grid Integration Systems – Energy Storage (SEGIS-ES)**" by *Dan Ton, Georgianne H Peek, Charles Hanley* and *John Boyes,* which outlines how the Solar Energy Grid Integration System (SEGIS) was initiated in early 2008 as industry led initiative to develop PV invertors, controllers, and energy management systems that will greatly enhance the utility of distributed PV systems. This was a result of the studies related to potential high penetration of distributed PV generation systems in the US national grid in late 2007 by the US Department of Energy. It describes the concept of augmenting the SEGIS with energy storage in residential and small commercial applications to increase penetration of distributed PV systems and prepare for grid ties reliability that is imminently required. Targeted research on applications for benefit from peak shaving, load shifting, demand response, outage protection and microgrids and development of optimized systems is in progress. The article describes the scope of the SEGIS-ES outlining the need to integrate energy storage with PV systems, applications for which it is most suited and for which it will provide the greatest economic and operational benefits to customers and utilities. And as selection and optimizing of storage technology for the application is critical to the success of any PV storage system, various technologies and comparative relative costs and development status is also given with identification of areas where further PV specific R&D is needed with procedural recommendations.

"**Solarscapes – A New Face for PV**" by *Topher Donahue,* is the next article which describes Solarscaping as nothing but landscaping in and around buildings with BIPV (building integrated photovoltaic). The idea of integrating PV into architectural structures is not new, the innovation is to offer a fresh design strategy to minimize the aesthetic concerns of adding PV to homes. PVs can be successfully integrated into awnings, pergolas, carports, hot tub shades, and gazebos. Even fences, fountains, greenhouses, archways and sunrooms are all good candidates – for a price. Historical homes with deep pocket owners are good candidates. The two basic elements of frame and bifacial modules (heteroinjection with intrinsic thin layer) is spacecraft technology. Once the investment is in and the payback starts, home owners are known to have seen their utility meters spin backward. The challenges are tall trees on the property. Traditional garden landscapers and solarscapers could be at loggerheads on this one, but solutions for array siting where shade free location is unavailable, with suitable integrating system design, location and site albedo are available. Attractive options of multifunctional architecture with range and sweep for heritage home, homesteads, farms abound along with prefabricated solar kits for building contractors for apartment blocks and townships.

"**Hybrid Solar Lighting Illuminates Energy Savings for Government Facilities**", works out how hybrid solar lighting provides an exciting new means of reducing energy consumption while also delivering significant benefits associated with natural lighting in commercial buildings. Hybrid lighting technology was originally developed for fluorescent lighting applications but was enhanced to work with incandescent accent-lighting commonly used in retail spaces where inefficient spending can total up to 55-70 per cent of energy allocations for lighting energy costs. Hybrid lighting has the potential to reduce energy consumption while also maintaining or exceeding high quality lighting requirements. This is seen as energy-efficient, higher quality, economically viable alternative to incandescent lamps. When the consideration for use of artificial lighting by industry is up to 25 per cent of the energy consumed in commercial buildings and industry, implementation of hybrid solar can significantly impact energy costs favourably. Many commercial buildings, industrial buildings, institutional facilities like schools, libraries and hospitals stand to benefit. R&D efforts are on to enhance the performance and reliability of the technology as well as the application of the system to work with newly emerging solid-state lighting sources to be even more flexible, convenient, reliable and control friendly.

The "**Solar Greenhouse Resources: Horticulture Resource List**" by *Barbara Bellows* and *Katherine Adam* is more than just that. It is also a literature survey of great practical use. It starts with the 2000 phenomenon of adoption of high tunnels as the preferred greenhouse technology in the US while rigid frames and glazing are still common in Europe, Mexico and the Caribbean mainly for producing acres of winter crops for North American markets. A discussion outlines the uses of the principle of passive solar design, hazards, due to weather turbulence, and optimum greenhouse design including orientation, glazing, insulation, storage, storage material, handling phase change and more. An active method for solar heating greenhouses uses subterranean heating or earth thermal storage heating. This is explained with good greenhouse management practices for best heat retrieval. Ventilation, which is the key to a successful greenhouse, is discussed outlining how active solar cooling systems include solar air conditioning units and PVs set up to run evaporative cooling pads. How incorporation of new glazing, heat storage and insulating materials into the design can greatly enhance the efficiencies of greenhouse structures are explained.

The third section titled **Experiences and Initiatives** has four articles.

The case on "**Electricity from Solar Home Systems in South Africa**" by *Gisela Prasad* discusses how developed countries have introduced Renewable Energy (RE) technologies to reduce Green House Gas emissions and developing countries have an economic agenda—access to electricity for the poor in remote and rural areas. The South African government generally supports RE and has a policy stipulating a voluntary target of 10,000 GWh to be supplied by renewable sources by 2013 which is approximately 10 per cent of the country's total electricity demand. This paper presents two case studies describing solar water heaters and electricity from solar home systems, both of which include the impact of poverty on the dissemination and acceptance of technology. Solar Home Systems were planned for remote and poor rural areas to which the Grids have not been or cannot be extended as a substitute for grid electricity; they have been slow and limited. Pursuing economic growth and environmental protection in a transition country like SA is a huge challenge as it has one of the cheapest electricities generated from very large coal deposits. Eventually, the poorest of the poor got excluded due to their inability to repay for a technology source that is abundant and free for all human beings. Yet the programme achieved relative success by its embedment into the local environment.

The next article titled "**Power for Africa: Solar Hybrid Systems for Rural Electrification**", outlines how developing countries and Third World countries are the losers of accelerated climate change, mindless of where it occurs and who caused it. These countries that rely largely on agriculture are severely affected by changing, extreme weather conditions. Since these countries do not have the money to adapt accordingly, the increasing scarceness and rising prices of fossil fuel hamper their development. Against this background, rural electrification with reasonable renewable energy becomes particularly important. Decentrally generated electricity allows access to clean water, to healthcare services, education and economic development. An estimated 1.5 billion people across the world must still live without electricity. In many regions without distant electricity, diesel for generation is prohibitive, and Jatropha plantations and solar installations are replacing oil imports. These have become technically possible and economically feasible. The example of the Good Vincent Sisters in Mbinga, Tanzania who use sunlight and vegetable oil for power generation is illustrated where a prize winning German firm developed this solar hybrid technology for them.

The article on "**Creating a Credit Market for Solar Thermal: The PROSOL Project in Tunisia**", by *Emanuela Menichetti* and *Myriem Touhami* is about a UNEP project under study that is part of the Mediterranean Renewable Energy Programme which works on expanding the share of renewable energy technologies in Southern Mediterranean region in Tunisia, Morocco and Egypt. The objective is to reduce poverty, combat climate change and achieve long-term sustainability objectives. This case looks at the Prosol project in Tunisia which has significant solar potential with very high irradiation rates and scant fossil fuels. The main features of the Prosol financing scheme are loan mechanism for domestic consumers to purchase solar water heaters, a governmental capital cost subsidy and discounted interest rates on the loans, progressively phased out. A series of accompanying measures include an awareness raising campaign, a capacity building programme and carbon finance. Measures have also been taken to create a level playing field where solar thermal can better compete with conventional energy sources. It shows how banks were also engaged in leveraging enough financial resources to stimulate the creation of a solar thermal market. Limited budget and money channelled in the proper direction with synergies created and exploited can create real success stories.

The last article titled "**Solar Power in India: The Nascent Stage**", by *Mohan Kumar Devarajan* discusses the solar status in India which has the world's largest programme for renewable energy and a huge need to reach its remote and far flung areas where conventional grid is not economical or cost-effective. Solar energy is being pursued through the thermal route and solar photovoltaic route. The article traces the growth of solar energy installations in India. A fine example of a Himalayan project is given to illustrate the intense need and its amazing impact of solar energy in ice-bound regions. The vital role of NGOs, collaborations with other countries along with financing options are discussed. It has tremendous market opportunity as 65 per cent of the country's population lives in the rural areas. These citizens need access to a better quality of life with interventions on a self managerial scale of costs.

Conclusion

Among renewable energy sources, solar power becomes the energy source of choice where consumers do not have the option of an electricity grid in the vicinity. Whether paneled or towered or built into the design of dwellings and factories, the solar option generally needs large flat open spaces and could be seen as landscape altering. Remote regions that were so far deprived of power and energy can join the mainstream economy in terms of facilities and progress. The business community needs to enlarge the size of the solar market and be energized by the possibilities of tomorrow. Clean and non-disruptive technologies like solar need to be supported and information about them disseminated.

SECTION I

VALUE PROPOSITION

1

Solar: The Most Important Renewable of the Future

Andrew Blakers

In this article, the author explains the importance of solar energy in the long-term. He emphasizes on the importance of increased support for renewable energy and energy efficiency R&D, demonstration, commercialisation, market incentives and technology incentives.

Summary

Climate change is a major issue. Energy efficiency, renewable energy, carbon capture and storage ("clean coal"), afforestation and other measures, all need to be harnessed to solve the problem. None of these options will prosper without a price on carbon pollution.

Renewable energy comprises many energy forms, including photovoltaics, solar thermal electricity, solar heat, wind, geothermal, bio energy, hydro, ocean energy, solar buildings and clean transport systems. Taken together, renewable energy and energy efficiency can provide most of Australia's energy needs by 2050. Renewable energy and energy efficiency are more job-intensive than coal fired power stations.

Source: http://solar.anu.edu.au © Andrew Blakers. Reprinted with permission. This article was originally published as "Sustainable Energy".

Direct solar energy such as photovoltaics and solar thermal, with a resource 1,000 larger than current energy consumption, is the most important of the renewable energy technologies in the long-term.

Baseload power can be met by wide geographical dispersion of collectors, technical diversity (using many different forms of renewable energy), storage (e.g. pumped hydro-pumping water uphill during the day and releasing it through turbines at night) and shifting loads from nighttime to daytime.

The worldwide solar energy industry is doubling in size every 20 months, and will reach $100 billion per year by about 2011. Wind energy is enjoying similar growth. Australia has real innovative strength in some sectors, notably photovoltaics. Australia can play an important environmental, employment and economic role in this industry.

It is important for Australia to have a balanced portfolio. Increased support is needed for renewable energy and energy efficiency R&D, demonstration, commercialisation and market incentives. The latter could include an extension to the Government's mandatory renewable energy target, carbon pricing and individual technology incentives.

Energy Supply Options

There are five available energy sources. These are energy from the sun (in its various forms), nuclear energy, fossil energy, tidal energy and geothermal energy.

Solar energy is available on a massive scale. The collection and conversion methods usually (but not always) entail few environmental or social problems. Solar energy includes both direct radiation and indirect forms such as biomass, wind, hydro, ocean thermal, ocean currents and waves. Most of these energy forms will be part of the energy mix when solar energy becomes the dominant traded-energy form.

Tidal energy can be collected using what amounts to a coastal hydroelectric system. It is sustainable in the sense that it will not run out. However, the coastline is a scarce resource and the collection of large amounts of tidal energy will have a major environmental impact.

Geothermal energy has its origins in the decay of radioactive elements within the Earth. Heat associated with volcanic regions can be used to generate steam for district heating or to drive a steam turbine to produce electricity. Another form is "hot dry rocks", which refers to hot masses of slightly radioactive rock buried several kilometres below the surface of the Earth. Cold water can be forced down to this rock, which is then fractured. Steam can be extracted from another borehole nearby. Geothermal energy is restricted to particular geographical locations. It is sustainable in the sense that it can be harvested with limited environmental damage, although the heat stored in a particular place can certainly be depleted.

Nuclear energy from fission has severe problems relating to nuclear weapons proliferation and nuclear terrorism. A country that possesses a nuclear energy industry has the raw materials, the technology and the trained people required for the production of nuclear weapons. Several countries have acquired nuclear weapons technology via the route of civilian nuclear energy. Other problems include waste disposal and reactor accidents. Nuclear fusion is still several decades away from commercial utilisation, but may make a major contribution to sustainable energy supply in the future.

Fossil fuels are the principal cause of the enhanced greenhouse effect and are subject to resource depletion. Other problems include oil spills, oil-related warfare (for example, the Gulf wars) and pollution from acid rain, particulates and photochemical smog.

The Enhanced Greenhouse Effect

There is a consensus among climate scientists that the burning of fossil fuels is causing an enhanced greenhouse effect [IPCC; Climate models]. Consequences over the next 50 years could include:

- significant temperature rises (particularly at high latitudes)
- rising sea temperatures & levels (causing flooding, coastal erosion, damage to coral reefs)
- more frequent extreme weather events (such as floods, storms and drought)

- the need to move agricultural activities and infrastructure to different locations
- an expanded range for tropical diseases and disease vectors
- substantial reduction in biodiversity and
- more severe bushfire seasons.

It is possible that there will be catastrophes over the next few decades such as positive feedbacks in climate change from large scale release of methane, loss of the Gulf Stream, due to changes in the circulation of the Atlantic Ocean, and destruction of the Amazon rainforest.

Solar Energy Options

Solar energy can eliminate the need for fossil and nuclear fuels over the next 50 years. Some solar energy technologies are more advanced than others. The key to successful mass-utilisation of solar energy is diversity. The solar energy mix will vary from region to region. Unlike fossil fuels, the resource is ubiquitous, which removes questions of monopolisation of energy supply.

Photovoltaics, solar heat and wind energy are the renewable energy technologies that can provide very large quantities of sustainable energy with high (more than 10 per cent) overall efficiency in order to limit land use requirements [Blakers 2000]. These conversion technologies have small environmental impacts and insignificant military or terrorist applications. In some countries biomass may also make a substantial contribution to energy supply, despite low conversion efficiency.

Indirect Solar Energy: Solar energy in the form of waves, ocean thermal gradients, ocean currents and hydro sources are geographically limited. Hydro energy is usually associated with large environmental impacts arising from the drowning of river valleys and the alteration of river hydrology. These energy forms could contribute substantially in particular regions to an environmentally responsible energy supply.

Biomass Energy: Biomass can be derived from waste materials such as sugar cane bagasse, garbage, sawdust and sewerage. Firewood is another common form of biomass energy. When biomass production for energy is combined with other useful purposes the economic viability can be substantially improved. For example, the growth of Eucalyptus Mallee trees in the corridors between fields allows the harvesting of wood by coppicing. The Mallee also serves other purposes including as a windbreak, shade and an ecological corridor. In a country such as Australia, where the population density is low, multiple-purpose energy crops may be viable. In developing countries, where energy use per capita is low, biomass can be a substantial fraction of total energy use.

Unfortunately, the conversion of solar energy into chemical biomass energy has a very low overall efficiency. The conversion of solar energy to chemical energy has an efficiency of less than one per cent, and conversion to electricity has efficiency below 0.5 per cent (which is two orders of magnitude smaller than from photovoltaics, solar heat or wind energy). When used on a very large scale, biomass competes with food and timber production or with ecosystem preservation for the supply of arable land, water, pesticides and fertiliser. The notion that biomass can be grown at low cost on waste land with small environmental impact is incorrect. The use of low quality agricultural land results in low yields and high costs, whether the crop is for food, timber or for energy. Suggestions that genetic engineering can be used to greatly increase the energy conversion efficiency of biomass are far from current reality.

Low Temperature Solar Heat: Good building design, which allows the use of natural solar heat and light, together with good insulation, minimises the requirement for space heating. Solar water heaters are directly competitive with electricity or gas in most parts of the world. Solar concentrator water heaters have been combined with photovoltaic collectors to produce 60–70 per cent efficient hot water and electricity systems.

Wind Energy: Modern wind generators have capacities in the multi-megawatt range. They have 40-70 metre high tubular steel towers on a concrete foundation and have three blades, each 20-40 metre long. They are computer-controlled and centrally monitored, with many safety features. They have availabilities above 98 per cent and will last more than 20 years. Wind generators alienate less land per unit of energy produced than any other energy source. Most windfarms are located on cleared farming land. Farming activities are scarcely affected, and the wind generators amount to a second cash crop for the farmer. A large fraction of future wind farms will be located offshore to take advantage of higher wind speeds and to avoid land use conflicts. Provided that a windfarm is not located in an ecologically sensitive area, the only significant environmental impact of wind energy is visual.

Wind energy is likely to generate 10-20 per cent of the world's electricity by 2030, and is now regarded as a conventional energy source [DWTA; EWEA]. The industry is growing at a rate of 20-30 per cent per year and has an annual turnover of around A$30 billion per year.

Photovoltaics: Photovoltaics (PV) is an elegant but expensive technology. It has found widespread use in niche markets such as consumer electronics, remote area power supplies and satellites. Large numbers of PV systems are being installed on house roofs in cities. The cost of PV systems is not a strong function of scale, which means that PV systems are often the most economical energy source for small applications. Over 90 per cent of the world photovoltaic market is serviced by crystalline silicon solar cells. This dominance is likely to continue for many years.

The worldwide photovoltaics (PV) industry has been doubling every 20 months for the past 7 years, which is far in excess of the growth rate in energy consumption. In 2006 the value of PV system sales was about $25 billion. Rapid growth in production is causing steady reductions in cost ("learning curve" reductions), which will eventually lead to a true mass-market developing.

The maximum theoretical efficiency of PV is 86%, compared with the current world record efficiency of 43%. Many decades will be required before PV technology approaches theoretical limits, and so the cost of PV systems can be confidently expected to continue to decline for decades – as has happened with the related integrated circuit industry.

Solar Thermal Electricity: Most solar thermal electricity technologies use mirrors to concentrate sunlight onto a receiver. The resulting heat is ultimately used to generate steam, which passes through a turbine to produce electricity. Two non-concentrating exceptions are solar chimneys and solar ponds. Concentrator methods are equally applicable to concentrating PV systems. The usual ways of concentrating sunlight are point focus concentrators (dishes), line focus concentrators (troughs, both reflective and refractive) and central receivers (heliostats and power towers).

Solar thermal electricity is not yet a commercial proposition. The reason for this is that, unlike PV, there are strong economies of scale. This means that small systems that might be suitable for an individual household are far too expensive. This lack of a niche market, in contrast to PV, inhibits the development of solar thermal electricity in the short to medium term.

The future of high temperature solar thermal might lie in the generation of thermochemicals and in the storage of heat at high temperature to allow for 24 hour power production.

High Temperature Solar Heat: Concentrated solar energy can achieve the same temperatures as fossil and nuclear fuels, either directly (using mirrors) or through the use of chemicals (thermochemicals or bio fuels) created using solar energy. One problem for high temperature solar heat is that heavy industry (e.g. the steel industry) is often located near coalfields, in regions that are relatively poorly endowed with solar energy. Perhaps the next steel mill could be built in the Pilbara region of Western Australia, close to iron ore deposits, rather than on the east coast, close to coal deposits. North west Australia is one of the sunniest places on earth.

Solar heat can be used to extend fossil fuels. For example, if natural gas (CH_4) is heated using a dish concentrator in the presence of steam (H_2O) then hydrogen and carbon monoxide ($3H_2 + CO$) are produced. The energy content of the hydrogen and carbon monoxide is about 30 per cent larger than that of the methane, and so solar energy has been added to the original methane.

Energy Efficiency: Hand in hand with the utilisation of solar energy goes energy efficiency. 'Solar energy' and 'energy efficiency' are often the same thing. For example, an energy-efficient building is a building that utilises natural solar light and heat sensibly. Walking rather than driving uses a small amount of solar energy (food) rather than a larger amount of oil energy. A clothesline, solar salt production and putting on extra clothing displaces an electric clothes dryer, fossil-fuel fired kiln drying of salt and gas or electric heating respectively.

Energy Issues

Energy "Payback" Time: The time required to recover the energy investment in solar energy equipment is typically one tenth of the lifetime of the equipment. The energy intensity and cost of solar systems are closely linked. Both are falling. Concentrator and thin-film PV systems, which are the technologies that will dominate after 2010, are likely to have energy payback times of about 1-2 years compared with a system lifetime of 30 years.

Energy Storage and Intermittancy: Energy storage issues are not likely to prove to be major obstacles to mass utilisation of solar energy. However, considerable work will be required on storage over the next four decades. Solar energy is generally less dispatchable than energy supply from fossil fuels. However, fossil fuel generation is not completely reliable. The question to be considered is how to ensure that the probability of failure of energy supply in a solar-dominated energy mix is similarly small to that in a fossil fuel dominated energy mix.

Biomass is dispatchable, in the sense that operators of a national grid can choose when to burn the biomass. Hot dry rocks and hydro have similar dispatachability to fossil fuel generators.

There are many options for the storage of low temperature solar heat (for water and space heating) that are simple and cheap. Examples include in building materials, water, crushed rock and phase-change materials. The latter are materials that melt and freeze at a particular temperature (e.g. 31C°). The large amount of heat energy liberated when a material freezes effectively pins the temperature of a house with phase change energy storage near the freezing point of the phase change material.

Another method of solar energy storage is via thermochemistry. One example that is being worked on at the Australian National University is ammonia [CSES]. Ammonia can be disassociated into hydrogen and nitrogen at the focus of a dish solar concentrator system. These gases can be pumped long distances in natural gas pipelines and recombined when convenient to form ammonia and to yield high temperature steam that is suitable for industrial use or electricity generation.

Storage of renewable energy electricity is not a serious issue until penetration of the grid reaches 10-20 per cent. This will not happen (except in a few places) for many years, giving time for improved storage technologies to be developed. A number of storage options are available, including thermochemical energy storage, fly wheels, compressed air and pumped hydro. The latter is a fully commercial technology that involves pumping water uphill to a reservoir during times of low power demand. The water is released through a turbine during periods of high demand. A river is not actually needed, since water can go around the cycle from upper to lower reservoir and return an indefinite number of times. For example, seawater and some local hills is sufficient.

Wide geographical dispersal minimises storage requirements. If the wind or sun is not available in one region then it might well be available in another region. The use of satellites and other means allows for a high degree of predictability for unavailability of solar or wind energy, which allows alternative generators to be brought on-line.

Technology diversity (the use of a variety of energy conversion technologies) minimises storage requirements. For example, the probability that neither wind energy nor photovoltaic energy will be available at a particular time is lower than the probability that either is not available. The probability is low that a combination of solar, wind, hydro, biomass, geothermal, ocean, tidal and judicious use of fossil fuels cannot meet a particular load.

If one particular renewable energy technology, such as photovoltaics, is a large component of annual energy supply, then the fact that the sun does not shine at night can be accommodated for by shifting loads to daytime where possible (for example, by time-of-day energy pricing).

Climate scientists believe that on-going use of fossil fuels at a rate of about 20% of current use is compatible with stabilisation of CO_2 emissions. Reservation of this tranche of energy supply for night time use (when solar is not available) and for peaking (e.g. from fast-response natural gas) allows considerable flexibility in managing a national grid.

In summary, geographical dispersal, technical diversity, the use of dispatchable technologies such as biomass, hot dry rocks and hydro, energy storage and the judicious use of relatively small quantities of natural gas will allow renewable energy to dominate electricity production while at the same time reducing fossil fuel use by 80%.

Hydrogen: A great deal has been written about the hydrogen economy. Hydrogen is a gas with a very low boiling point (-253° C). There are several technical problems associated with hydrogen, including the difficulty of storage. Hydrogen storage could be by way of compression or liquefaction, or by reversible sorption processes or chemical reactions. Storage for use in cars using methods that could compete economically with petrol or compressed natural gas is a challenging proposition, but is receiving considerable attention.

By far the largest problem is the question of the source of the hydrogen. Hydrogen is currently derived by reacting water with fossil fuels at high temperatures. This produces carbon dioxide and is no more sustainable than any other fossil fuel burning technology. Electricity can be used to split water into hydrogen and oxygen (electrolysis). In principle, the electricity can be from a sustainable solar energy source. Unfortunately, solar electricity is expensive. Electrolysis of water to produce hydrogen entails substantial conversion losses, and conversion of the energy in hydrogen to any form of energy other than heat entails further substantial losses. For example, the round trip efficiency of renewable electricity followed by electrolysis to form hydrogen followed by conversion of hydrogen back to electricity using a fuel cell is less than 50 per cent, which is much lower than other storage techniques. In most cases it is cheaper and more efficient to use the renewable electricity directly. The direct splitting of water under sunlight (e.g. by using titanium dioxide) has formidable technical obstacles, relating to corrosion and very low conversion efficiency, which are far from resolution.

Fuel Cells: Fuel cells convert gaseous fuels to electricity without combustion. They have the potential to be considerably more efficient than conventional combustion, particularly in small systems. Substantial technical obstacles still remain. Fuel cells are sometimes claimed to be a renewable energy enabling technology because they could convert hydrogen energy (produced using solar energy) to electricity at relatively high efficiencies. However, there are too many conversion losses in this sequence, which amounts to a storage technology. Other storage technologies are usually cheaper, more efficient and more practical.

Fuel cells may have important applications in saving energy. For example, the use of fuel cells in houses to produce electricity from natural gas, with space and water heating as a by-product, could considerably reduce total greenhouse gas emissions from houses, particularly in cold climates.

Solar Energy for Transport: Liquid fuels from biomass (e.g. ethanol) can power vehicles, but only at substantial environmental cost if used on a large scale. The private car in cities can be largely replaced by public transport, which is much more energy efficient than a car and can be powered by renewable electricity. Freight can be shifted to electrically powered trains. Lightweight electric cars are more efficient than current automobiles for city use. It is an open question as to whether significant private motor vehicle ownership can be afforded in an environmentally-constrained world. Imagine India and China having similar car ownership levels to Australia.

Carbon Sequestration: The fossil fuel industry is devoting considerable resources to the development of methods of storing carbon dioxide from power stations underground or in the ocean. One method is to separate the carbon dioxide from products of combustion, compress it and pump it into saline aquifers. Current estimates for the cost that 'zero emission coal' electricity might reach over the next decade or two are comparable with estimates for various forms of solar electricity. Carbon sequestration may assist the transition to a greenhouse-neutral energy economy in the period to 2050.

Large Scale Utilisation of Solar Energy for Electricity Production

Photovoltaics, solar heat and wind energy, together with modest contributions from other sources, can reduce consumption of fossil and nuclear fuels by 80% or more. The limited wind energy resources of the world means that wind will probably only contribute about 25% of the world's electricity in the long-term.

The efficiency of solar thermal electric and solar photovoltaic energy systems will converge over the next decade to 15-25%. That is, at noon on a sunny day the electricity production will be about 200 Watts per square metre of solar collecting area. Allowance must be made for the fact that the collectors are spaced apart to avoid self-shading and that solar power systems only operate during sunlight hours. Taking these factors and all other losses into account, average power densities (24 hours per day) will be about 10-12 W/m^2 or 100 Gigawatt hours (GWh) per year per square km of land.

The proportion of the Earth's land surface that must be set to provide all of the world's current electricity consumption is about 0.1%, much of it in arid sunny regions. If solar energy also provides solar heat and solar chemicals to replace all fossil fuel use then two to three times as much land would be required. It is clear that land availability is not a constraint on a 100% solar energy future.

Approximately 2,600 km^2 of land (0.035% of Australia's land surface area) would be needed for PV panels to replace all of Australia's electricity from fossil fuels. This assumes that only one third of the land is covered by PV panels (to avoid self-shading). This is a tiny fraction of the area occupied by Australian cities. To put this in perspective, each person in Australia would need to have access to $40m^2$ of PV panel to eliminate the need to burn fossil fuels. An average suburban house block is $500\text{-}1{,}000m^2$. The area of house roof in Australia is sufficient to replace all of Australia's electricity from fossil fuels. Thus the area of land alienated by mass utilisation of solar energy is relatively very small.

In order to achieve carbon neutrality, a panel area of about $30m^2$ would be needed for a 4-5 star (energy rating) house with gas space heating, solar water heating and efficient electrical appliances. With such a panel the house would export as much electricity to the grid during the day as it imports at night. An additional $10m^2$ pf PV panel would be required to offset the greenhouse gas emissions from a modern car travelling 10,000-12,000 km/year. An additional $5m^2$ of PV panel would be required to offset the greenhouse gas emissions from a gas space heating system. A solar water heater of $5m^2$ would complete the solar collector system. The $50m^2$ of roof area required to make a house greenhouse neutral is much smaller than the typical roof area of $150\text{-}200m^2$.

PV and solar thermal energy collection alienates 3-10 times less land than does hydroelectricity per unit of annual energy production (and nil in the case of solar collectors on building roofs). It is comparable to the area of land alienated by fossil and nuclear power stations over their lifetimes, including the power stations themselves and their security perimeters, mines, transport corridors, waste dumps, cooling lakes and other infrastructure. Wind energy alienates very little land per megawatt, with farming operations continuing all around the turbine towers.

Material usage for solar energy conversion is small. The principle elements used in a PV system are extremely abundant on earth: silicon, oxygen, hydrogen, carbon, iron and calcium, with small amounts of other elements also being used. PV technology entails far lower need for mining (a factor of between 100 and 1,000 in the amount of material that is mined per unit of energy produced) than does nuclear or coal energy. Interestingly, for an advanced thin film silicon PV technology such as Sliver, the energy production per kg of silicon over the lifetime of the PV system is the same as the energy production per kg of refined nuclear fuel in a reactor.

Environmental Impacts

The area alienated by the towers of a wind farm is less than 1% of the area spanned by the wind farm. Wind farms amount to a second cash crop for the farmer, with farming operations continuing largely unaffected around the towers. Increasingly, windfarms will be offshore and will therefore not alienate any land at all.

Solar water heaters, solar air heaters and photovoltaic solar electric systems located on building roofs alienate no land at all. There is enough roof space on Australian homes to supply all of Australia's electricity from solar energy.

Gram for gram, advanced thin film silicon solar cells produce the same amount of electricity over their lifetime as nuclear fuel rods. Per tonne of mined material, solar and wind energy systems have vastly better lifetime energy yield than either nuclear or fossil energy systems.

The principal elements required for solar and wind energy systems (silicon, oxygen, hydrogen, carbon, sodium, potassium, calcium, aluminium and iron) are among the most abundant on earth. Carbon dioxide emissions per unit of useful energy produced are among the lowest of all energy systems, and continue to decline. Only 0.1% of the world's land surface area would be required to supply all of the world's electricity from solar and wind energy—mostly in arid regions and on building roofs. Solar and wind energy can supply most of the world's energy needs with small environmental cost.

Cost of Solar Energy

The fact that there is currently no penalty for greenhouse gas emissions means that clean energy technologies have difficulty competing with fossil fuels. When the cost of geosequestration of carbon dioxide is added in to the cost of electricity from fossil fuels, the generation cost will rise to around A$0.07-0.10/kWh, which will cause the retail price of electricity in a low cost country like the USA or Australia to rise by about a quarter.

Wind energy can already compete with the generation cost of "zero emission coal". On current trends, solar thermal electricity and photovoltaics will also compete successfully within 20 years, and perhaps sooner. Low temperature solar thermal energy for water and space heating is already economic in most places. The provision of transport fuel and of high temperature heat for industry are more difficult challenges than the provision of solar electricity and low temperature solar heat. The challenge can be met by a combination of energy conservation and shifting loads to electricity (e.g. electric cars and trains).

Renewable Energy in Australia

Australia has excellent and diverse solar energy resources by world standards. Australia has a low population density, is one of the sunniest countries in the world and has vast wind resources, particularly along the southern coastline. Since it is a large country with a small population, biomass can make a major contribution. Australia also has large uranium, coal, oil and gas deposits.

The Australian Government has committed large resources to strategic fossil fuel R&D through a variety of mechanisms, including several Cooperative Research Centres, several CSIRO Divisions, the Rio Tinto Foundation, Geoscience Australia and others. Carbon sequestration is a particular focus. Such support is scarcely available to renewable energy. It is highly desirable for the Government to broaden its climate change portfolio to include support on a similar scale for renewable energy.

Photovoltaics and wind energy are likely to be A$100 billion per year industries by 2011. Australian Government policy over the past decade has been less supportive of the renewable energy industry than in comparable countries such as Germany and Japan. Nevertheless, Australia retains a foothold in the renewable energy industry, and silicon photovoltaic cells in particular. BP Solar and Rheem/ Solahart have major manufacturing facilities in Australia for photovoltaics and solar water heaters respectively. Several wind energy companies are active in Australia. Two Australian companies (Origin Energy and Solar Systems) are taking new PV technologies to market.

Universities have been powerhouses for renewable energy innovation in Australia. Australian university renewable energy technology is highly valued in international technology markets. In particular, photovoltaics is an area of real Australian research and commercialisation strength. Two universities retain critical mass (more than 10 people engaged in R&D) in the area of solar energy: the Australian National University [CSES] and the University of NSW [KCPVE]. It is important to nurture University-based R&D, both directly for the development of new technologies and to service the research & training needs of a rapidly growing local industry. The new $150 million renewable energy innovation fund is a good start, but will need to greatly expanded over the next few years to be commensurate with the opportunities and the scale of the climate change problem.

The Mandatory Renewable Energy Target of the new Government requires that Australian electricity companies produce an additional 45000-60000 GWh of renewable electricity by 2020. The MRET program is an excellent market support mechanism. In effect, a very small tax per kWh on the vast electricity industry is used to support the nascent renewable energy industry. In order to achieve its potential it is desirable that ambitious MRET targets be set for each year out to 2050. MRET could be expanded to include other low emission technologies, including, importantly, energy efficiency.

Unfortunately other support mechanisms for renewable energy are weak compared with our foreign competitors. In particular, support mechanisms that favour manufacturers based in Australia need to be strengthened. The new $500 million renewable energy innovation fund could provide a significant boost to local manufacturing if the rules are drafted appropriately.

Support for renewable energy in Australia should preferably be focused on IP generation and the export of IP-rich high-value products and services. Australia is not a low cost manufacturing nation. Fortunately, Australia has substantial expertise in several renewable energy technologies, notably photovoltaics, that lend themselves well to such a strategy. This strategy would comprise substantial support for R&D, company-University interaction and professional education in Universities and CSIRO, coupled with strong incentives for companies to manufacture high value products in Australia for export and to license IP overseas.

Future Retail Cost of Electricity in Australia

A typical domestic retail cost of electricity in Australia is 14 c/kWh. Regardless of future carbon pollution pricing, this is likely to increase over time because the cost of carbon fuel is increasing in line with the cost of oil.

Carbon pollution pricing is likely to add 2 c/kWh in the short-term and 5 c/kWh in the longer term to the retail price of electricity. The introduction of time-of-day pricing will favour technologies that produce electricity at times of peak demand, which is typically on summer afternoons driven by air conditioning. This is likely to confer a price advantage of 5 c/kWh or more for technologies such as photovoltaics and solar thermal power. Mandatory low emission technology targets for 2020, arising from promises made in the 2007 Federal election, will confer a price advantage of about 5 c/kWh on renewable energy technologies.

Based on the above analysis, photovoltaic systems on house roofs, competing at the retail level, will be a fully competitive option if they can produce electricity for less than 30 c/kWh. This condition is met if the fully installed net cost (after all subsidies) of the photovoltaic system is $5,000 per nominal kW. This calculation assumes an interest/discount rate of 8% (approximating the mortgage rate), a system life of 30 years, replacement of the inverter after 10 years, solar insolation typical of SE Australia, and a system capacity factor of 15% (realistically taking account of losses such as dirty glass, shading from nearby trees and buildings, elevated cell temperature loss, inversion losses and a 1% per year decline in system output as the system ages). Such an installed system cost can be reached with a quite modest subsidy in large scale installation of PV systems; for example, installing PV on every house in a new suburb. In the not too distant future no subsidy will be required.

Conclusion

Wind energy, solar thermal and photovoltaics are the only truly large-scale renewable energy generation technologies available. They are each likely to be $100 billion/year industries within 5-8 years. These technologies are relatively free of adverse environmental impacts, and have the potential to dominate the traded energy market over the next 50 years.

It is likely that international concern over the enhanced greenhouse effect will continue to increase. The consequence of this concern will be ever increasing support for solar energy around the world. It is to be hoped that Australian government policies will be such as to place Australian companies in the forefront of this rapidly growing industry.

Dedicated and strategically directed funding of solar energy R&D and research and professional training, together with reliable long-term commercialisation support for Australian-based manufacturing, is required if Australia is to become a major player in this vast new industry.

(Andrew Blakers is Director of Centre for Sustainable Energy Systems at Australian National University. The author can be reached at Andrew.blakers@anu.edu.au).

References and General Reading

Blakers, 'Solar and Wind Electricity in Australia', *Australian Journal of Environmental Management*, Vol 7, pp 223-236, 2000, *<http://solar.anu.edu.au>*

CSES. Centre for Sustainable Energy Systems, Australian National University: *<http://solar.anu.edu.au>*

Climate models: *<http://www.dar.csiro.au/publications/projections2001.pdf>*

DWTA. Danish Wind Turbine Manufacturers Association: *<http://www.windpower.dk>*

EWEA. European Wind Energy Association: *<http://www.ewea.org>*

Geodynamics: *<http://www.geodynamics.com.au/IRM/content>*

IPCC. Intergovernmental Panel on Climate Change: *<http://www.ipcc.ch>*

KCPVE. Key Centre for PV Engineering, University of NSW: *<http://www.pv.unsw.edu.au>*

2

Payback and other Financial Tests for Solar Electric Systems

Andy Black

The article focuses on residential analysis for homes in Northern California's PG&E utility territory. It explains the factors that improve payback for solar electric systems and also how to determine the payback. The author has shown the increase in resale value of properties installed with solar electric systems.

How to Calculate the Return on your Solar Electric System Investment before you Buy.

For years, questions about returns on the expensive investment in a solar electric system were dismissed with the analogy, "What's the payback on your swimming pool?" That sentiment might speak to the converted, but for most people considering a solar energy system, the financial case is a major deciding factor.

Fortunately, photovoltaic (PV) technology has matured such that the payback question can now be given a serious answer, backed by solid math and accounting. The answers vary significantly by local climate, utility rates and incentives.

Source: www.ongrid.net © 2007, Andy Black. Reprinted with permission.

In the best cases in California, the compound annual rate of return is well over 10 percent, the cash flow is positive, and the increase in property resale value more than covers the cost of the PV system. In other parts of the country where electric rates are low and incentives may be less, a grid-tied system may barely cover its maintenance costs.

This article focuses on residential analyses for homes in Northern California's PG&E utility territory. Similar calculations can be done for commercial situations, but significant differences in the tax and accounting rules exist.

What Factors Improve Payback?

The most important factors for making solar an attractive investment include high electric rates, financial incentives, net-metering policies and good sunlight (available in almost all of the continental United States).

High electric rates can come in various ways. California has a tiered pricing system penalizing large residential users with prices as high as $0.36 per kilowatt-hour (see Figure 1).

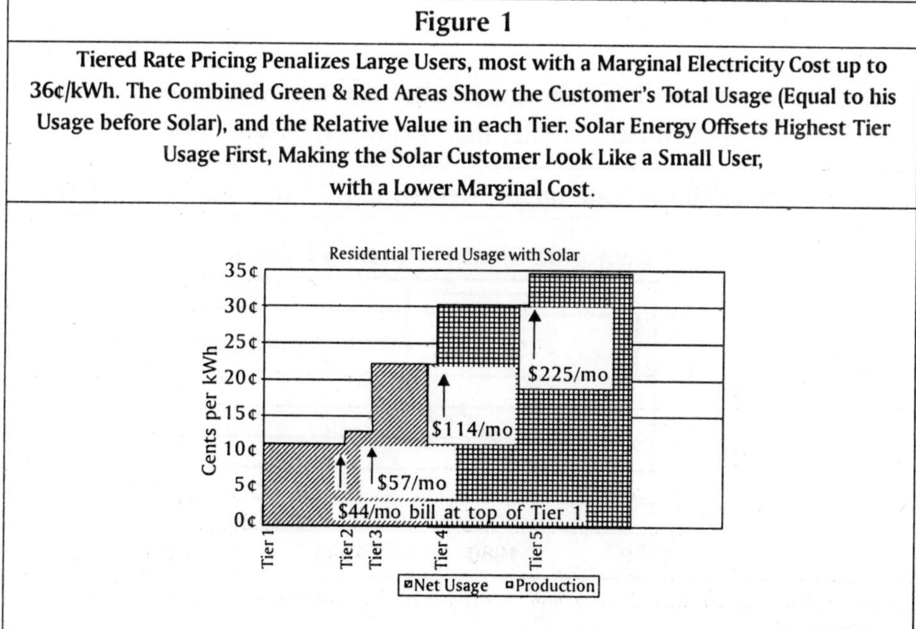

Figure 1

Tiered Rate Pricing Penalizes Large Users, most with a Marginal Electricity Cost up to 36¢/kWh. The Combined Green & Red Areas Show the Customer's Total Usage (Equal to his Usage before Solar), and the Relative Value in each Tier. Solar Energy Offsets Highest Tier Usage First, Making the Solar Customer Look Like a Small User, with a Lower Marginal Cost.

Under net-metering laws, solar energy offsets the retail cost of the electricity generated. Even better, in California, solar systems are allowed to operate on a time-of-use rate schedule, which enables users to sell back electricity to the utility at peak rates, which can be even more valuable. Time-of-use rates vary electricity price by time of day, with the higher rates occurring during times of shortage, when the utility must pay more to purchase electricity from generators. Solar tends to produce its electricity during these higher rate periods. These high rates are the most important factor in improving the payback.

Direct incentives can include tax benefits such as credits or depreciation. The most celebrated recent incentive is the federal tax credit for solar systems that on went into effect on January 1, 2006. The credit is for 30 percent of the system cost up to $2,000 for residential systems (no cap on commercial credits). For PV systems, that typically means a $2,000 credit on the purchaser's tax return for the year the system was installed. This credit can be coupled with state, local & utility incentives. One such popular incentive is the California Solar Initiative (CSI) rebate, which can discount up to an additional 30-40 percent of the system's cost. Consult a certified tax advisor to check the applicability of such incentives to your situation.

A big factor in the economics is inflation in electric rates (see Figure 2). Solar is an inflation-protected investment, because it offsets electricity costs at the current prevailing retail rate. As rates rise, the owner saves even more.

Figure 2: Rates have Gone up an Average of 6.7% per Year for 30 Years

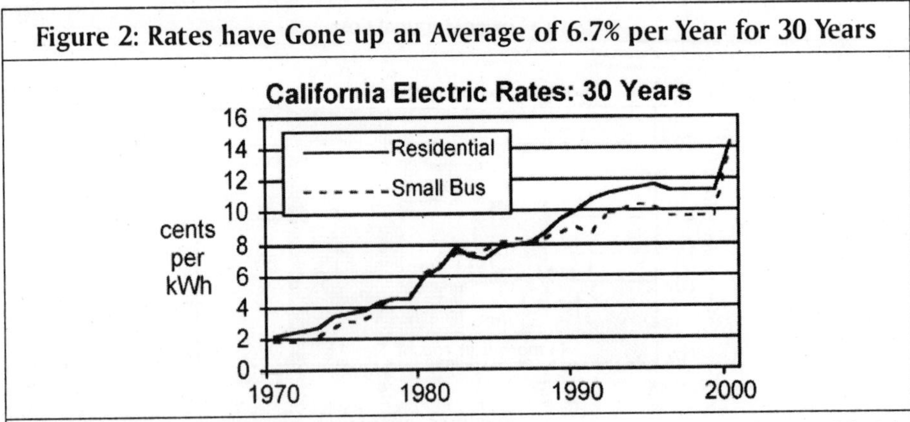

Source: CPUC "Electric Rate Compendium" November 2001. This article assumes inflation will be 5% going forward.

Newer forms of direct incentives are PBIs (performance based incentives) and RECs (renewable energy credits, or "green tags"). Both incentives are paid on a per kilowatt-hour basis. Unlike rebates, they don't help reduce the up-front cost, but they do increase the cash payments received after commissioning the system. Payments can be as much as $0.39 per kilowatt-hour for five years for the PBI, and between $.01 than $0.05 per kilowatt-hour for a 5-plus-year contract on RECs if the system size is large enough (usually at least 10kW). Because these payments often can be combined with net metering value, the PV system is capable of garnering substantial revenue per kilowatt-hour generated.

Determining the Payback

Several useful ways to measure the economic value of a solar system are: compound annual rate of return, cash flow and increase in property resale value. In strong economic cases, the annual returns will be over 10 percent, the cash flow positive and the increase in resale value greater than system cost. These are common returns in PG&E territory.

Compound annual rate of return, or CARR, on an investment is another term for interest-rate yield, which is a way of comparing one investment to another. For example, a savings account might pay 1 percent interest and the long-term stock market has paid about 10.5 percent. Solar systems in California can often see a pre-tax CARR of 10 percent or more.

Several examples are shown in Table 1. (For more detailed information on these and the subsequent calculations, see the articles at *www.ongrid.net/papers*)

The cash flow will be positive, either immediately or within the first few years, for many homeowners who finance their solar systems using home equity loans.

The cash flow calculation compares the estimated savings on the electric bill to the cost of the loan. Monthly loan cost is the principal plus interest payment required to pay off the loan, less any tax savings. In the case of "deductible" loans, such as home equity-based loans, the interest is usually tax-deductible and thus the loan effectively costs less. Home equity loans are also excellent sources of funds because interest rates on real estate-secured loans are relatively low and payment terms can be long.

Table 1: Example Residential Systems and their Financial Costs and Benefits

Before Solar		Solar System Size & Cost				Results, Savings, & Benefits					
Electric Bill	kWh Usage per month	PV System Size (CEC Rating)	Gross Cost	PG&E Rebate @ $2.43/W	Final Net Cost w/$ 2k Fed Tax Credit	Pre-Tax Compound Annual Rate of Return	Appraisal Equity/ Resale Increase in First Year	New Electric Bill with Solar	Net Monthly Cash Flow Compared to 7.5% 30-yr Loan		Total Savings Over 25 Years
									In First Year	In Fifth Year	
$ 100	675	3.0 kW	$28K	$7K	$20K	11.3%	$20K	$11/mo	$-13/mo	$0/mo	$43K
$ 200	1010	5.0 kW	$46K	$12K	$33K	14.5%	$44K	$6/mo	$26/mo	$62/mo	$96K
$ 374	1500	8.0 kW	$72K	$19K	$51K	17.1%	$84K	$7/mo	$106/mo	$181/mo	$183K
$ 374	1500	5.0 kW	$46K	$12K	$33K	19.9%	$63K	$102/mo	$104/mo	$160/mo	$137K

Inflation plays an important part. Inflation affects electric rates and thus effectively increases the savings from a solar system over time. Inflation doesn't affect loan rates, particularly with fixed-rate loans. Hence, as electric rates rise, the savings grows, but the cost of the loan stays relatively constant (it rises a little over time as the interest portion of the payment declines and the tax deductibility declines). See Figure 3 for an example.

Figure 3: Inflation's Effect on Loan Costs Versus Electric Costs without Solar

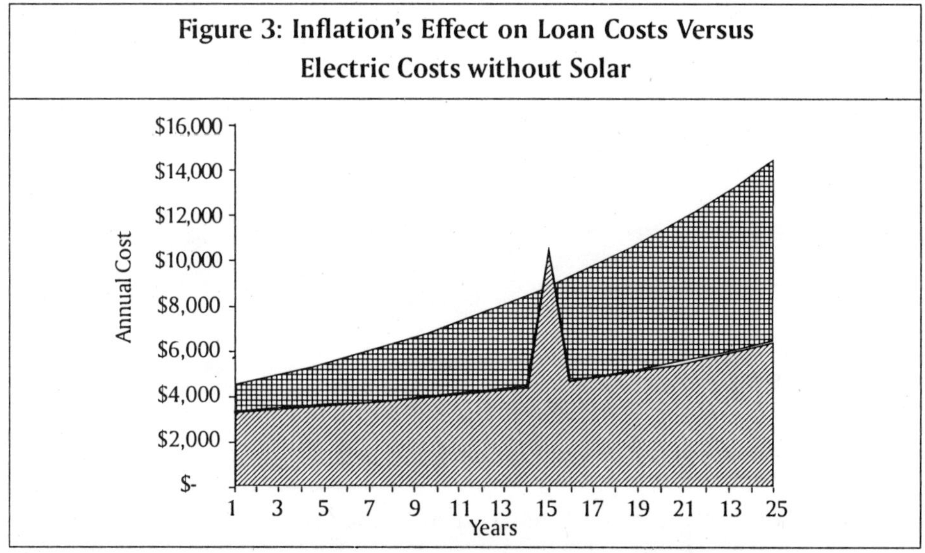

Figure 4 highlights the difference in the curves in Figure 3. This figure shows the net annual savings—old bill minus new bill, loan, maintenance and inverter-replacement costs—and shows the effects of inflation over time.

Figure 5 shows the accumulation of net annual savings. This accumulation is free and clear with no initial outlay of cash, because that was covered by the loan. In the example in Figure 5, the system will save the owner about $90,000 over 25 years, with no up-front cost. The savings are small though significant in the first years, but increase over time due to inflation. Of course, the purchaser can select any loan term that suits his needs.

Table 1 includes several examples showing the initial monthly cash flow, assuming 100 percent financing of a solar system's final net cost.

Figure 4: Net Annual Savings with a Photovoltaic System Net annual savings consists of the old electric bill minus the new bill, loan principal payments, loan net interest (after taxes), maintenance and inverter-placement costs

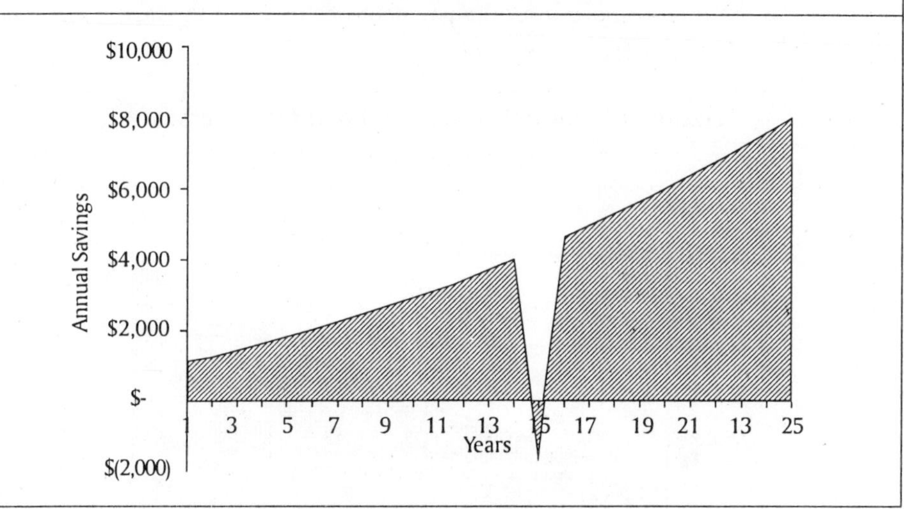

Figure 5: Accumulated Savings with a Photovoltaic System Accumulated Savings over 25 years with no "Out of Pocket" Costs to the Purchaser

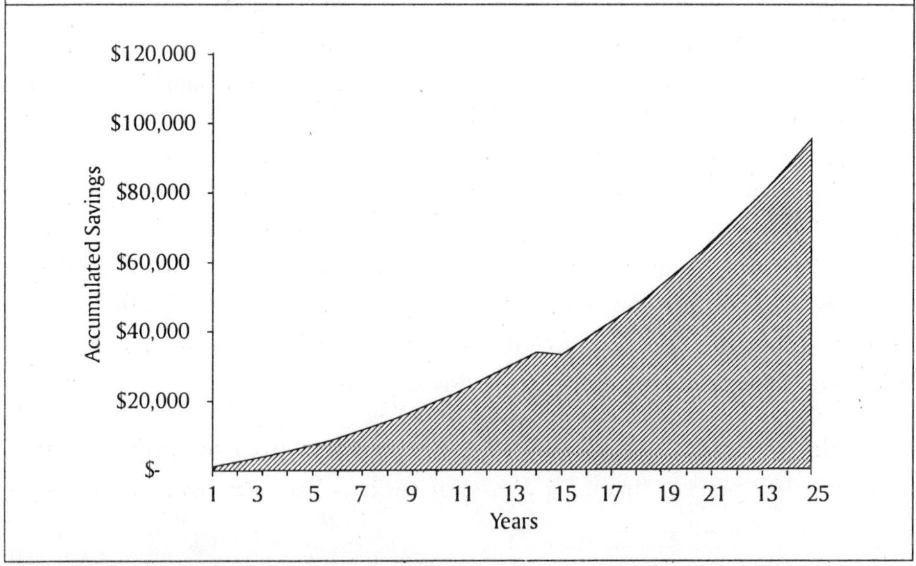

An Increase in Property Resale Value occurs in homes with solar electric systems because these systems decrease utility operating costs. According to a 1998 *Appraisal Journal* article by Rick Nevin and Gregory Watson, a home's value increases $20,000 for every $1,000 reduction in annual operating costs from energy efficiency. *(See www.icfconsulting.com/Markets/Community_Development/doc_files/apj1098.pdf)*

The rationale is that the money from the reduction in operating costs can be spent on a larger mortgage with no net change in monthly cost of ownership. Nevin and Watson state that historic mortgage costs have an after-tax effective interest rate of about 5 percent. If $1,000 of reduced operating costs is put toward debt service at 5 percent, it can support an additional $20,000 of debt. To the borrower, total monthly cost of home ownership is identical. Instead of paying the utility, the homeowner pays the bank, but the total cost is unchanged.

See the column labeled "Appraisal Equity Increase" in the Table for examples of the increase in home value. This increase can effectively reduce the payback period to zero years if the owner chose to sell the property immediately, and it removes the purchase risk. It could even lead to a profit on resale in some cases.

This increase in property value to date is currently theoretical. A high fraction of the grid-tied solar electric systems have been installed since 2001. Most of these homes have not been sold, so there are no broad studies of comparable resale values available. However, emerging evidence suggests that some solar-home sellers are realizing significant jumps in resale value.

A 2004 National Renewable Energy Laboratory (NREL) study demonstrated that San Diego zero-energy homes with solar features increased in value faster than comparable conventional homes in a nearby community. *(Access the report at www.nrel.gov/docs/fy04osti/35912.pdf.)* On average, the homes increased in value $40,000 more than the conventional homes, at a higher rate of appreciation and with a shorter length of ownership.

PV systems will appreciate over time, rather than depreciate as they age. The appreciation comes from the increasing annual savings the system will yield as electric rates and bill savings rise. All the calculations in this article assume annual electric rate inflation will be 5 percent. If so, the PV system will save 5 percent more value each successive year, and thus gain from the 20:1 multiplier effect. The property resale value will then increase 5 percent per year compounded.

This appreciation cannot continue forever, as the increase in resale value runs into the second limit, which relates to the system's remaining life. For these analyses, the system is assumed to be worthless at the end of 25 years. This estimate is conservative, since the panels are warranted at 25 years to work at 80 percent of their new capability. If the system is worthless at the end of 25 years, the only value the system has as it nears that time are the savings it can generate before the end of the 25th year. Figure 6 shows both the increasing property resale value due to increasing annual savings and the remaining-value limitation that takes over at approximately year 11. As the NREL resale study suggests, however, actual resale could be much higher depending on the market mood for solar.

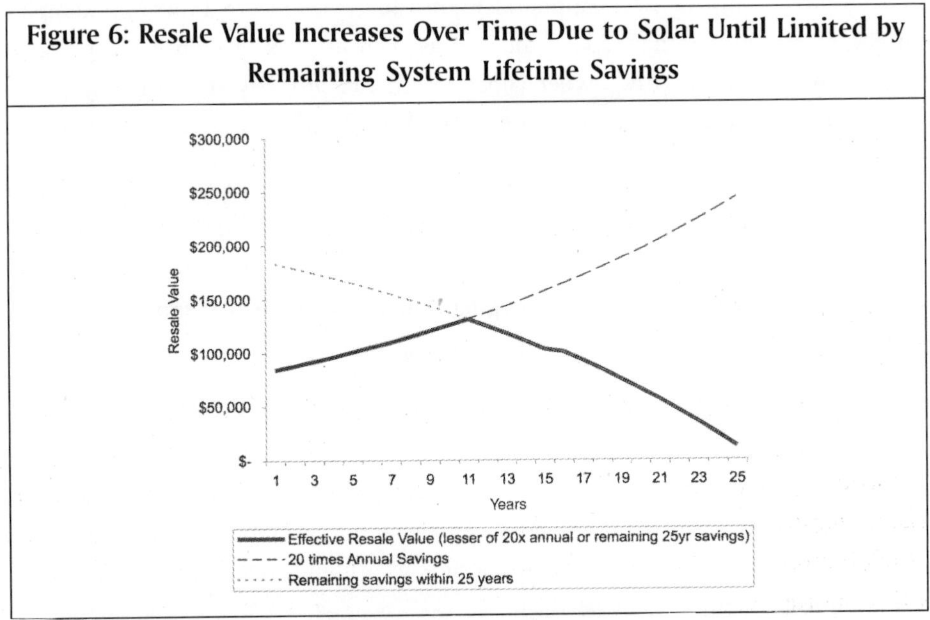

Figure 6: Resale Value Increases Over Time Due to Solar Until Limited by Remaining System Lifetime Savings

Creating Markets That Reward Investment

For more information on calculating photovoltaic system economic analyses and to see the analyses in this article explained in more depth, access *www.ongrid.net/papers*.

Be aware that some tools don't account for tiered or time-of-use electric rates in interaction with PV production, and as such, their results may over- or underestimate the value of a PV system in a particular situation.

Solar has finally come into its own in certain markets. These markets are exploding because individuals are discovering the financial benefits of owning PV systems in those regions. In order to encourage widespread adoption of solar energy, we need to empower everyone with this knowledge and expand the components that make it possible—tiered rates, time-of-use net metering, RPSs with solar requirements and RECs. Market forces will take it from there.

(Andy Black is CEO of OnGrid Solar, providing solar financial analysis tools and consultation. He serves as a board member of the American Solar Energy Society and the Advisory Board of NorCal Solar. Contact him at 408.428.0808 or www.ongrid.net. Black regularly teaches and consults on "Payback on Solar Electricity" for many audiences. The author can be reached at andy@ongrid.net).

3

Exploring the Economic Value of EPAct 2005's PV Tax Credits

Mark Bolinger, Ryan Wiser and Edwin Ing

The article examines the value and explores the implications of photovoltaic tax credits using a generic cash flow model. It emphasizes on interaction between governmental credits and grants for different types of entities, like purchasers and administrators. The findings also have important implications for policy design, and the type of incentives offered to residential and commercial system owners.

Introduction

The market for grid-connected photovoltaics (PV) in the US has grown dramatically in recent years, driven in large part by PV grant or "buy-down" programs in California, New Jersey, and many other states. The recent announcement of a new 11-year, $3.2 billion PV program in California suggests that state policy will continue to drive even faster growth over the next decade. Federal policy has also played a role, primarily by providing commercial PV systems access to tax benefits, including accelerated depreciation (5-year MACRS schedule) and a business energy Investment Tax Credit (ITC).

Source: http://eetd.lbl.gov/ea/ems/cases/LBNL_59928.pdf © The United States Government. Reprinted with permission.

Since the signing of the Energy Policy Act of 2005 (EPAct) on August 8, the federal government has begun to play a much more significant role in supporting both commercial and residential PV systems. Specifically, EPAct increased the federal ITC for commercial PV systems from 10% to 30% of system costs, and also created a new 30% ITC (capped at $2000) for residential solar systems. Both changes went into effect on January 1, 2006, for an initial period of two years, and in late 2006 were extended for an additional year. Unless extended further, the new residential ITC will expire, and the 30% commercial ITC will revert back to 10%, on January 1, 2009.

How much economic value do these new and expanded federal tax credits really provide to PV system purchasers? And what implications might they hold for state/utility PV grant programs? Using a generic (i.e., non-state-specific) cash flow model, this report explores these questions[1]. We begin with a discussion of the taxability of PV grants and their interaction with federal credits, as this issue significantly affects the analysis that follows. We then calculate the incremental value of EPAct's new and expanded credits for PV systems of different sizes, and owned by different types of entities. We conclude with a discussion of potential implications for purchasers of PV systems, as well as for administrators of state/utility PV programs.

Taxation of State Grants and Interaction with Federal Tax Credits

Perhaps surprisingly, whether or not the Internal Revenue Service (IRS) considers grants made by state/utility PV programs to be taxable income is a critical factor in determining the value of the new and expanded tax credits under EPAct[2]. This is because, at least for the foreseeable future, most PV systems in the US are likely to be installed with the financial support of a state/utility PV program. If the grants provided by these programs are considered to be taxable income, then a grant recipient can claim the federal ITC (and depreciation if a commercial system) on the full cost or "basis" of the system. If, however, the grants are *not* considered to be taxable income, then the grant recipient must reduce, by the amount of the grant, the basis to which the federal ITC (and depreciation) apply[3]. As a result, the taxable or non-taxable status of the grant carries significant federal tax consequences. PV grants frequently buy-down as much as 50% of installed system costs. If non-taxable, such grants will cut in half not only the value of the federal ITC, but also the tax benefits of depreciation (for commercial systems).

The economic impact is not trivial. For example, under the assumptions described later, our cash flow model reveals that a residential PV system garners the same value (in terms of net present value of after-tax cash flows) from a $2.7/W non-taxable grant as it does from a $4/W taxable grant; conversely, it would take a non-taxable grant of $5.8/W to provide a commercial system with the same after-tax value as a $4/W taxable grant.

This example demonstrates not only the magnitude of the impact, but also its disparate effect on residential and commercial PV systems. Because of the $2000 federal ITC cap for residential systems (which will be binding for all but the smallest systems) and the absence of depreciation benefits, residential systems are better off financially with a non-taxable, rather than taxable, grant. The opposite is true for commercial systems, which are better off paying income tax on the grant, and then applying the uncapped ITC and accelerated depreciation to the full basis of the project.

Given the degree of economic impact at stake, whether or not PV grants are taxable is clearly an important question[4]. Unfortunately, the IRS has provided limited direct guidance on this issue. Section 61 of the Internal Revenue Code generally defines gross (taxable) income to mean income derived from any source, except as otherwise provided in statute. The IRS broadly interprets this definition to treat government grants as taxable income unless statutorily excluded from taxation.

Though this suggests that PV grants should generally be considered taxable, four possible grounds for exclusion from taxable income might be explored. Specifically, a PV grant would be considered non-taxable if it were found to be one of the following: (1) a government social welfare payment; (2) a manufacturer or dealer rebate of the purchase price; (3) a contribution to the capital of a corporation; or (4) a utility energy conservation subsidy. Below we discuss each of these possibilities in turn, focusing in particular on what appears to be the most relevant potential exclusion – the possibility that a PV grant might qualify as a utility energy conservation subsidy[5].

Government Social Welfare Payments: While broadly defining taxable income to cover government grants, the IRS as a matter of public policy has created an exclusion for government welfare payments to individuals. In order to be non-taxable, however, such payments must be based on the recipient's established need. Since few if any PV programs require grant recipients to establish need, this exclusion is not particularly applicable (one possible exception might be PV grants offered specifically to low-income households).

Manufacturer or Dealer Rebate of the Purchase Price: Certain reductions in the purchase price of an asset may be considered non-taxable. In Technical Advice Memorandum 8924002, the IRS reviewed its past revenue rulings and concluded that "in order for the receipt of funds to be considered a non-taxable price rebate that reduces the basis of an item of property, several features must be present: (1) the rebate must be based on the purchase price of the item; (2) the manufacturer or dealer of the item must be the party offering the rebate; and (3) the recipient must be able to negotiate or renegotiate the purchase price in an arms-length transaction. ...therefore, a [non-taxable] rebate is treated as a reimbursement of the purchase price and not an accession to wealth."

Most PV programs fail to meet at least the first two requirements for this price rebate exclusion. Specifically, most programs base their grants on the size of the system (e.g., $/W), rather than on its purchase price. Furthermore, while in some instances state PV programs do provide grants to system retailers or installers, who in turn pass them through to system owners in the form of a reduced purchase price, in substance the grant is from the PV program (the retailer or installer would not have reduced the purchase price without having received the grant) and therefore the price rebate exclusion is not likely to apply[6].

Contribution to the Capital of a Corporation: Section 118 of the Internal Revenue Code excludes from taxable income contributions to the capital of a corporation. This exclusion applies to money transferred to a corporation (but not other types of businesses, such as LLCs or partnerships) by a government unit in order to obtain an advantage for the general community, rather than for direct services or recompense. Moreover, the contribution must, among other things: (1) become a permanent part of the recipient's working capital and not be used for paying dividends, interest, or anything else chargeable to or payable out of earnings or income; (2) be employed in or contribute to the production of additional income to the recipient; and (3) be bargained for by the recipient[7].

Since, in most cases, a PV grant recipient does not "bargain" for the grant[8], and is not obligated to use the grant in the manner specified above, this exclusion might be difficult to justify. Furthermore, given that corporations (as well as other commercial entities) are better off with a taxable PV grant, it is unclear why a corporation would ever even try to make the case that PV grants should qualify for the Section 118 exclusion from taxation. Instead, a more conservative (and, incidentally, lucrative) approach would be to simply assume that grants are taxable – which, after all, is the default position taken by the IRS under Section 61 (unless otherwise provided in statute).

Since we have heard anecdotally, however, that at least some (or perhaps even many) corporations in California have, in fact, treated PV grants as contributions to capital, we allow for this possibility in the analysis presented later in this paper.

Utility Energy Conservation Subsidy: Since 1991, Section 136 of the Internal Revenue Code has treated certain utility energy conservation subsidies as non-taxable income. Specifically, Section 136(a) states that "Gross income shall not include the value of any subsidy provided (directly or indirectly) by a public utility to a customer for the purchase or installation of any energy conservation measure." Section 136(c)(1) defines the term "energy conservation measure" to mean "any installation or modification primarily designed to reduce consumption of electricity or natural gas or to improve the management of energy demand with respect to a dwelling unit." This definition covers some solar energy systems, including PV systems[9].

A key question relating to this exclusion is exactly what is meant by "provided (directly or indirectly) by a public utility." Section 136(c)(2)(b) defines the term "public utility" to mean "a person engaged in the sale of electricity or natural gas to residential, commercial, or industrial customers for use by such customers. For purposes of the preceding sentence, the term "person" includes the Federal Government, a State or local government or any political subdivision thereof, or any instrumentality of any of the foregoing." Clearly, the administrators of most PV programs in the US (excepting those administered by utilities) are not "engaged in the sale of electricity," and so do not directly qualify as a public utility.

But might such programs be considered to indirectly provide energy conservation subsidies from a public utility? In many instances, state renewable energy funds (the non-utility administrators of most PV programs in the US) are financed by utilities or their ratepayers, thereby raising the possibility that they are, in fact, indirectly providing energy conservation subsidies from a public utility. The conference report to the Energy Policy Act of 1992, however, indicates that Congress inserted the "directly or indirectly" phrasing in Section 136 to prevent third party contractors (e.g., equipment vendors, such as PV retailers or installers) from taking advantage of the Section 136 exclusion. Specifically, it states:

"...the conferees believe that third party contractors should not be at a competitive advantage or disadvantage with respect to the tax benefits provided by the exclusion. In addition, the conferees believe that when a utility provides a payment to a third party contractor, the utility is indirectly providing the subsidy to the person for whom the contractor is providing the energy conservation measure and the exclusion should apply to such person. Thus, the conference agreement provides that the exclusion applies to any subsidy provided directly or indirectly to a utility customer, if such subsidy otherwise would be included in income. For example, if a public utility provides a subsidy to a customer to partially offset the cost of the installation of an energy conservation measure on the customer's premises, the provision [Section 136] applies to exclude [from taxable income] all or a portion of the value of such subsidy. Likewise, if the public utility provides a payment to an independent contractor so that the contractor can provide for the installation of an energy conservation measure on the utility customer's premises at a reduced price, the [Section 136] exclusion applies to the customer for the indirect subsidy supplied to the customer." [Emphasis added.]

In other words, the conference report discusses the "direct or indirect" subsidy only in the context of independent contractors – e.g., equipment vendors that might otherwise gain "a competitive advantage" if provided access to the exclusion. It does not directly touch on the broader issues of program administration or governmental subsidies.

Earlier IRS revenue and private letter rulings – though on a different statutory provision – do address these broader issues, and in some cases could be interpreted as indicating that the source of a program's funds would characterize the program[10]. Subsequently in Private Letter Ruling 853004 (April 30, 1985), however, the IRS indicated that a subsidy administered by a governmental unit would be treated as a government program whatever the funding source, suggesting that utility-funded, government-administered programs would not qualify for the Section 136 exclusion. Furthermore, the congressional report on the subsequent enactment of the Section 136 exclusion for utility energy conservation subsidies contains no express repudiation of the IRS' previous position on government subsidies. One might, therefore, expect the IRS to stick to its position taken in Private Letter Ruling 853004 that characterizes a government-administered program as a government program, regardless of the funding source.

Nevertheless, some uncertainty remains over the scope of the exclusion provided under Section 136 as it relates to state PV programs. Specifically, some of the rulings cited above – which, it should be noted, concerned credits and statutory provisions somewhat different from those of interest here – conflict with one another, and only address this issue peripherally (i.e., in commentary not necessary to the legal holding of the case). Furthermore, these rulings distinguish primarily between utility- and government-administered programs, raising the question of how the IRS might characterize programs funded by utilities (or their ratepayers) but administered by non-utility, non-governmental entities (e.g., non-profit administrators, such as the Energy Trust of Oregon or the Sustainable Development Fund in southeastern Pennsylvania).

New insight on this question arrived in January 2007, when the IRS found (in a private letter ruling) that the Energy Trust of Oregon's residential PV incentives do qualify for the Section 136 exclusion from gross income. The ruling seems to rest on the Energy Trust – a 501(c)(3) non-profit – being a non-governmental entity that indirectly provides the utility-funded (i.e., ratepayer-funded) subsidy. In aggregate, payments made by the Energy Trust to residential customers of a particular utility do not exceed funds collected from that utility, thereby helping to substantiate the program as a utility program that benefits that utility's customers.

With respect to the "directly or indirectly" question, the Oregon ruling simply states that "The legislative history of [Section 136] clarifies that the subsidy need not be provided directly by the public utility to the customer, and that the exclusion [from gross income] applies to the customer to whom a subsidy may be indirectly provided by the utility." This statement seems to indicate a broader IRS interpretation of the conference report to the Energy Policy Act of 1992 than previously presented in this report. That is, it seems to indicate that the IRS does not consider the "directly or indirectly" language to refer only to third-party contractors or equipment vendors who might otherwise benefit (in lieu of the utility's customers) from the Section 136 exclusion. Without further textual clarity, however, a definitive interpretation of the IRS's position on this issue is not possible[11].

In short, the IRS rulings to date on Section 136 suggest that otherwise-qualifying subsidy payments provided under a utility-administered program will qualify for the exclusion; those provided under a government-administered program likely will not qualify for the exclusion (regardless of the funding source)[12]; and those administered by a non-profit might qualify if the program can be characterized as a utility program. Given, however, a degree of lingering uncertainty, and that private letter rulings may not be used or cited as precedent, individual PV programs seeking clarity on this issue may wish to consult directly with the IRS.

Summary: In summary, though it is difficult to generalize, given the highly factual nature of the law surrounding this issue, it appears that grants made to commercial PV systems will, in most cases, likely not qualify for any of the four exclusions discussed above, and will therefore be considered taxable grants that do not reduce the project's basis to which the federal ITC and depreciation applies. The one potential (though perhaps unlikely) exception would be if PV grants to corporations were determined by the IRS to be contributions to capital under Section 118, in which case corporations – but not other types of businesses, such as partnerships or LLCs – would need to exclude the grants from gross income, and reduce the project's tax basis by the amount of the grant. The taxability of grants made to residential PV systems will vary based on whether those grants are administered (either directly or indirectly) as a utility program under Section 136, with some uncertainty as to what what exactly constitutes a "utility" program[13]. To reflect this outstanding uncertainty, our analysis below allows for the possibility of either taxable or non-taxable grants to both residential and commercial systems.

Analysis

To examine the potential value of EPAct's new and expanded PV tax credits, we developed a cash flow model of a PV system in a generic state that offers a buy-down grant (either taxable or non-taxable) of $4/W.[14] Our approach was to determine how much this $4/W grant could be reduced, given EPAct's new or expanded PV tax credits, such that the PV system purchaser would remain indifferent (between pre- and post-EPAct conditions) in terms of the net present value of after-tax cash flows. The size of the reduction can be thought of not only as the maximum amount by which a PV program could reduce the size of its grants without causing the after-tax economics of PV to deteriorate (relative to pre-EPAct conditions), but also as the maximum value of the EPAct credits, on a grant-equivalent, $/W basis.

The resulting values for systems sized between 1 and 20 kW are shown in Figure 1. For 1 kW residential systems, the new EPAct ITC provides the same value as a non-taxable grant of $1.9/W (or a taxable grant of $2.7/W).[15] This value, however, drops precipitously to around $0.5/W non-taxable (or $0.7/W taxable) for 4 kW systems, and to $0.2/W non-taxable (or $0.3 taxable) at 10 kW. This decay in value as system size increases is due to the $2000 cap on the credit, which contributes an increasingly smaller proportion of total costs as system size increases. Because the 30% commercial ITC is not similarly capped, its value (relative to the 10% ITC available previously) remains fairly constant across different system sizes (even much larger system sizes than shown – e.g., 250 kW), equivalent to a taxable grant of just over $2.00/W (or non-taxable grant of just over $1.50/W). Finally, EPAct's PV tax credits provide no value to tax-exempt entities, to those subject to the alternative minimum tax, or to entities with no tax liability for other reasons.

Interestingly, with the exception of non-taxable commercial grants, our results are not dependent on our baseline assumption of a $4/W grant (vs. a $2/W grant, for example). In the case of a taxable grant, the size of the grant is immaterial (at least for this purpose), as it does not reduce the project's basis, and therefore does not impact the incremental value of the ITC or depreciation. In the case of a residential non-taxable grant, the ITC is almost always (except for the smallest systems) capped at $2000 – even after reducing the project's basis by the grant amount – so again the size of the grant that we have assumed is, for the most part, immaterial to our results. Commercial non-taxable grants, however, will impact the size of the uncapped ITC, meaning that our results for this special (though perhaps unlikely) case will vary depending on the grant size assumed.

Discussion

Results of the analysis presented above hold important implications for both PV system purchasers and administrators of PV programs. For PV system purchasers, it is clear from Figure 1 that the economic value of EPAct's new and expanded tax credits is strongly dependent on system size as well as the type and tax status of the system owner. Commercial PV system owners with tax liability will benefit greatly from the expanded ITC, as will owners of small residential systems from the new residential ITC. On the other hand, larger residential systems and systems owned by entities with limited or no tax liability (e.g., municipalities, non-profits) will gain little from the EPAct credits. These differences will no doubt affect the nature of consumer demand for PV while the credits are in effect: home-owners may demand smaller PV systems, while larger entities with limited or no tax liability may increasingly choose third-party ownership to indirectly capture the benefits of these new credits.

At the same time, EPAct's credits may not ultimately be worth as much as the maximum values presented in Figure 1,[16] because PV programs could (and in some cases have done so) reclaim some or all of EPAct's value (while still leaving system purchasers no worse off than before EPAct) by reducing the size of grants offered. Reducing grant size can help to stretch program budgets over a larger number of PV installations, without unduly suppressing growth in the market. Furthermore, by targeting any reductions at those specific system sizes and types that stand to benefit the most from EPAct – e.g., commercial and small residential systems – program administrators may help to level the playing field and ensure that EPAct does not favor certain market segments (e.g., commercial and small residential systems) while disadvantaging others (e.g., tax-exempt and large residential systems).

Reducing grant size – even in a targetted fashion – by the maximum amounts represented in Figure 1, however, may not be ideal for a number of reasons. Worldwide demand for solar modules and the increase in the cost of silicon feedstock have pushed PV module costs higher in recent years. Program administrators may wish to let the new and expanded federal credits offset this price increase, and perhaps even go a bit further to boost return on investment and thereby stimulate additional demand for PV. Furthermore, EPAct's tax credits may not be perceived by consumers to be as valuable as a grant that reduces up-front cash outlays. Finally, unless extended, EPAct's new and expanded PV tax credits will expire at the end of 2008.

In part as a result of these factors, the Solar Energy Industries Association (SEIA) has recommended that any reduction in rebate levels not exceed 50% of the estimated value of the Federal ITC[17]. To date, several PV programs, including those in New Jersey, Oregon, and Wisconsin, have reduced their grant levels in response to EPAct's new and expanded federal credits. For the most part, these reductions have been relatively modest (except perhaps with respect to larger residential systems) in comparison to the maximum incremental value of the EPAct credits presented in Figure 1.

Figure 1: Incremental Value of EPAct PV Tax Credits

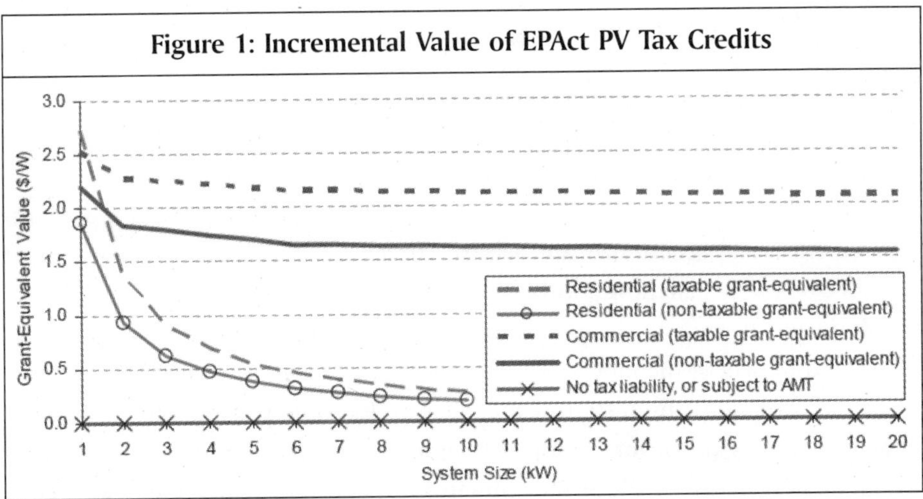

Our findings also have important implications for policy design, and the type of incentive offered. Specifically, many PV programs (including California's new $3.2 billion, 11-year solar initiative) are considering shifting (or have shifted) from capacity-based incentives (i.e., the $/W grants described in this paper) to performance-based incentives (i.e., $/kWh payments over time), at least in part based on the belief that capacity-based incentives reduce the project's basis to which federal tax credits and depreciation apply, making them less valuable than performance-based incentives, which do not reduce basis. As shown in this report, however, capacity-based incentives will only reduce tax basis if they are non-taxable, which in many cases appears unlikely (particularly for commercial systems) given the tax analysis presented above. As such, though there may be good reasons for shifting to performance-based incentives, maximizing the value of federal tax credits may not be among them.

Finally, the fact that residential system owners benefit most from non-taxable grants (while commercial owners prefer taxable grants) has implications for PV program administration. Where possible, new PV incentive programs for residential customers would ideally be administered in a way so that non-taxable grants can be provided (e.g., by utilities or non-profits that fall under the Section 136 exclusion). Within existing programs, administrators may want to seek clarification from the IRS – perhaps using arguments that capitalize on some of the uncertainties presented in this report – that their residential (but not commercial) incentives are non-taxable.

Acknowledgement

Lawrence Berkeley National Laboratory's contribution to this article was funded by the Clean Energy States Alliance, the US Department of Energy (Office of Electricity Delivery and Energy Reliability, Permitting, Siting and Analysis) under Contract No. DE-AC02-05CH11231, and the National Renewable Energy Laboratory under Memorandum Purchase Order No.DEK-6-66278-01.

(Mark Bolinger is Energy Analyst at Lawrence Berkeley National Laboratory. The author can be reached at MABolinger@lbl.gov,

Ryan Wiser is a scientist in the Electricity Markets and Policy Group at Lawrence Berkeley National Laboratory. The author can be reached at RHWiser@lbl.gov, and

Edwin Ing, Law Offices of Edwin T.C. Ing.)

Endnotes

1 For an application of this model, and the concepts included in this report, to California, see Ryan Wiser and Mark Bolinger, "Federal Tax Incentives for PV: Potential Implications for Program Design" (*http://eetd.lbl.gov/ea/emp/reports Wiser_Bolinger_CPUC_PV_Tax_03_2006.pdf*)

2 For earlier work on this topic, see Susan Gouchoe, Lynne Gillette, Christy Herig. 2004. "Are Solar Rebates and Grants for Homeowners and Businesses Taxable?" ASES SOLAR 2004 Proceedings. *http://www.dsireusa.org/documents/PolicyPublications/Taxability_ASES_2004.pdf*

3 See the conference report to the Crude Oil Windfall Profits Tax Act of 1980, which states that "under present law...if property is financed with nontaxable government grants, the tax basis in the property, for such purposes as depreciation and investment credits (including energy investment credits), is reduced to the extent that the property is financed with such grants." It goes on to explain that "grants which are taxable are not taken into account under these [credit offset] rules because their taxation serves as a partial offset; similarly, credits against State and local income taxes are not taken into account because the deductibility of these taxes under the Federal income tax implies that the effect of these credits is equivalent to the effect of a taxable grant."

4 Though a useful and otherwise thorough reference document, the March 2006 Solar Energy Industries Association's (SEIA) "Guide to Federal Tax Incentives for Solar Energy (Version 1.1)" does not directly address this question. For more information, see *http://www.seia.org/manualdownload.php*.

5 The tax analysis that follows does not take into account, and may not be applicable to, third-party-owned PV systems.

6 It is also worth noting that taxation cannot be avoided by providing grants to retailers or installers, rather than system owners. As shown later in this section (in the discussion of the Section 136 exclusion), the IRS has clearly held that any tax liability (or exclusion from tax liability) associated with a grant rests with the intended recipient (in this case, the system owner), and cannot be shifted to the retailer, installer, or any other third party. Thus, in the event that retailers do pass through taxable grants to PV system purchasers, both parties are obligated to pay income tax on the grant amount (and the system purchaser must also increase the basis of the system to its full, undiscounted value). Though at first blush this may seem like double taxation, it is no different from the more straightforward case in which the grant goes to the system purchaser, who pays income tax on the grant (but does not reduce basis) and provides the retailer with taxable revenue equivalent to the full, undiscounted cost of the system.

7 General Counsel Memorandum 37354 (December 21, 1977); Private Letter Ruling 9401035 (October 14, 1993); Edwards v. Cuba Railroad Co (268 US 628, 1925); Detroit Edison Co. (319 US 98, 1943); US v. Chicago, Burlington & Quincy Railroad (412 US 401, 1973).

8 This bargaining requirement would, if taken literally, appear to be difficult to satisfy in the case of most government grants. Revenue Ruling 93-16 (1993-1 Cumulative Bulletin 26), however, addressed this requirement with respect to FAA grants to airport owners. In that ruling, the IRS deemed the grants to be "bargained for" because they were "competitive, highly sought after, and made pursuant to meaningful criteria and conditions" (Kimberly S. Blanchard, "The Taxability of Capital Subsidies and Other Targeted Incentives," Tax Notes, November 8, 1999).

9 Section 210(11) of the National Energy Conservation Policy Act of 1978 (Public Law 95-619) defines "residential energy conservation measures" to include "devices to utilize solar energy or windpower for any residential energy conservation purpose, including heating of water, space heating and cooling...that are warranted by the manufacturer to meet a specified level of performance over a period of not less than three years." The Energy Policy Act of 1992, which first implemented Section 136 of the tax code, appears to have adopted this definition (at least according to the conference report – the specific adoption or definition does not appear to be codified in the Act or in Section 136 of the code). Finally, the IRS recently found (through a private letter ruling) that the Energy Trust of Oregon's cash incentives for PV systems do qualify for the Section 136 exclusion, and therefore that PV is an eligible "energy conservation measure."

10 See Revenue Ruling 81-52 (1981-1 Cumulative Bulletin 9); Revenue Ruling 83-145 (1983-2 Cumulative Bulletin 14); and Private Letter Ruling 8342047 (July 18, 1983).

11 In arguing its case with respect to this "directly or indirectly" issue, the Energy Trust stressed the following: that it is a tax-exempt non-profit entity that was created specifically for the purpose of administering the utilities' conservation and renewable energy programs; that it does so through contractual and "fiduciary-like" relationships with the utilities; that its programs have replaced those previously offered by the utilities; and that it has brought a new-found predictability and stability to conservation and renewable energy programs, thereby benefiting the state of Oregon as well as ratepayers of the participating utilities. It is not clear which, if any, of these arguments influenced the IRS's conclusion that the Energy Trust indirectly provides the subsidy on behalf of the utilities.

12 The recent Oregon private letter ruling clouds the issue with respect to governmental administrators. Previous IRS rulings (discussed earlier) suggested that a government-administered program would never be considered a utility program, and therefore would not qualify for the Section 136 exclusion. The recent Oregon private letter ruling, however, suggests that the IRS will, in some cases, allow utility programs that are not administered by a utility to qualify for the exclusion. Hence, it follows that if a governmental administrator can make a strong case that it is administering a utility program, it is possible that the IRS might find the program eligible for the Section 136 exclusion.

13 Section 136 does not apply to commercial systems, and so cannot be used to argue for tax-exempt treatment of grants to such systems. Though Section 136 originally included – with limitations – commercial energy conservation measures as well, these were ultimately stripped out by the Small Business Job Protection Act of 1996.

14 Other assumptions include: a cash-financed system with a 25-year project life; some economies of scale in installed costs ($10/W at 1 kW, $9/W at 2 kW, $8.5/W at 6 kW, and $8.2/W at 20 kW, with linear interpolation between these points); 15.4% capacity factor (i.e., 1350 kWh/kW/year); $0.12/kWh avoided electricity cost, escalating at 3%/year (treated as taxable income for commercial, but not residential, systems); no state tax credits; state depreciation follows federal (i.e., 5-year MACRS); federal ITC reduces basis for federal depreciation by half of the ITC (i.e., 15%); federal ITC does not reduce basis for state depreciation; tax brackets of 28% (federal residential), 34% (federal commercial), and 8% (state residential and commercial); $4/W grant is either taxable or non-taxable at both the federal and state level; state income tax payments are deductible from federal income; and a 7% nominal discount rate.

15 The fact that the taxable grant-equivalent value is higher than the non-taxable grant-equivalent value should not be interpreted to mean that a residential system owner is better off with a taxable grant; indeed, as described earlier at the beginning of the tax analysis section, the opposite is true. Instead, this taxable/non-taxable differential is due to the fact that a taxable grant represents pre-tax income, whereas a non-taxable grant represents after-tax income. Reducing the size of a taxable grant (e.g., in response to EPAct) reduces the recipient's tax liability without impacting the value of the EPAct credit; this reduction in tax liability, in turn, allows a further reduction in grant size (a positive feedback) relative to a non-taxable grant.

16 It should be emphasized that the analysis presented in this paper is generic (i.e., not state-specific), and that outcomes will differ in individual states that offer state tax incentives, or present other complexities. State-specific analysis is required to determine the true value of EPAct tax credits under any specific PV program. Footnote 1, for example, provides a citation for analysis conducted specifically for California's PV program administrators.

17 In addition, SEIA recommends that the total program budget be maintained (i.e., so that the program is able to support a greater number PV systems at the reduced grant level), that any reductions in grant size be made in such a way as to not degrade the economics for any customer class (i.e., use differentiated incentives), and that changes be made in a transparent and forward-looking fashion.

4

Brokering for Reducing Costs and Increasing Customer Satisfaction in Solar Sales

Andy Black

The concept of the brokering model for sales of solar systems can result in cost savings and increased benefits and satisfaction for both the installer and customer through significant improvements in efficiencies in the sales process. Because of the increased value added, it can result in increased income to the broker professional as well. However, there are significantly increased risks for both the customer and installer. To mitigate these risks, greater requirements for the training, knowledge, and experience of the broker are necessary.

The author has developed a unique and successful business model based on full service brokering of solar systems and believes that it would significantly reduce sales costs for the PV industry, while increasing customer satisfaction.

Source: www.ongrid.net © 2007, Andy Black. Reprinted with permission.

1. Introduction

The concept of brokering the sale of a PV system is borrowed from other broker type sales industries such as real estate. The broker forms relationships with potential customers and with several installers of turnkey solar systems. The broker conducts a customer site assessment and analysis, designs a suitable PV system, and solicits bids for the design from several of the installers. The broker analyzes the bids, presents the bids and the bid analysis to the customer, and assists the customer in making a determination of which bid to accept based on their expert knowledge and local experience, knowing each of the installers and their backgrounds, plus any additional factors that may be relevant. If a sale is made, the broker receives a commission from the installer.

2. The Brokering Concept

The brokering idea is borrowed from many other industries that have transitioned from sales using a dedicated seller's agent who represents only one seller and that seller's interests, to sales using a middle agent who works with several sellers and several customers simultaneously. Perhaps the closest parallel is the real estate industry. Because the value of the item is large and the valuation complex, the customer is generally not expected to become an expert in all the nuances and subtleties, so an expert agent (broker) can be very useful.

The broker is usually paid on commission by the chosen installer. However, the commission rate is the same, regardless of chosen installer, so the broker has very little financial bias between systems (only a small bias due to the minor price variations). This allows the broker to be indifferent, and therefore a good assistant and consultant in helping the customer choose the best solution. Because of this indifference, the customer may choose to confide more in the broker about concerns or true motivations they might otherwise hide from the more self-interested salespeople.

3. Benefits from Brokering

There are a number of significant benefits with brokers over traditional selling strategies. These include increases in efficiency, reductions in costs, increases in satisfaction for both customer and installer, and installer margin protection.

3.1 Greater Overall Selling Efficiency

The broker brings significant gains in efficiency to the purchase of a solar system. Rather than two to five salespeople bidding and competing for a sale, each having to do a site visit, proposal, and presentation, involving 1 to 2 trips (each) to the site, the broker can accomplish the same work with one person at far less total effort and expense (sales expenses can be significant, such as mileage, printing, communication, overhead, benefits, etc).

Further still, a skilled and experienced broker can implement customer-screening techniques that reduce time spent on "Looky Lous" and other non-serious customers.

It is important to note that in some states, the broker may not legally charge a fee for their services if they also might receive a commission, because that creates a conflict of fiduciary responsibilities – only the real estate industry appears to be allowed to engage in "dual agencies". However, there are ways of screening customers and keeping the broker's time investment focused on the serious purchasers. The author has developed a number of screening techniques that may be presented in a future paper. One of the reasons a broker can be more successful in screening than a salesperson is that they are bringing the customer much more value, and can request something in return, whereas a salesperson is just another bidder, and has relatively little leverage to demand any level of seriousness from the customer.

In so doing the broker's relative time efficiency compared to the regular solar salesperson is dramatically increased. Commonly, a customer will get 3 bids, giving each salesperson only a 33% chance of closing the sale, if the customer actually purchases. However, a significant number of potential customers never purchase at all (50% or more), so in reality, a salesman's chances of closing are often fewer than 1 in 6 or less than 17%.

On the other hand, a broker with a good screening mechanism is both bringing all the bids, as well as avoiding the non-serious customers. As a result, the broker is likely to close a very high percentage. The author's experience has been above 75%, or about 5 times his average when he worked as a solar salesman dedicated to one company.

There is a little more work involved as a broker – the broker must write up the bid specification and then normalize the bids, but there is less “salesmanship” involved – the customer is presented with all their options and information, rather than a crafted presentation designed to influence their decision.

3.2 Cost Savings

The broker bears more costs. Unlike the salesperson, whose costs are usually covered at least in part by their company, the broker absorbs all of the sales costs and probably most of the marketing costs for the sales they bring. However, the dramatic increase in selling efficiency can result in several cost savings:

- The company no longer bears the sales expenses and marketing costs (of the broker sales), which can be an additional 3-6% of the sale price, over and above the commission costs. The broker’s sales and marketing costs per sale are dramatically lower because of the greater efficiency
- The company no longer needs to support as many inefficient sales staff with overhead, tools and benefits
- The broker may charge a slightly lower commission because they are doing a lot less work for each sale by working 4 to 5 times more efficiently. Supply and demand and competition will dictate what the brokers are able to charge for their services.

For example, a 6% commissioned sales rep having their sales expenses covered (mileage, printing, postage, phone, internet, overhead, benefits) can cost the company and additional 4% per sale (total = 10%), plus marketing expenses. A well compensated solar broker, whose total commission is 7% easily saves the company 3% or more. In competitive bidding, at least some of this savings is passed on to the customer; the balance is enjoyed as additional profit margin for the installer. And while the broker had to absorb the sales expenses, those were much less than the 4% because they make so many more sales per dollar spent. Therefore, their real sales costs were probably closer to 1%, so their net commission is the same 6% as the inside sales person, but they can close 4 to 5 times as many sales with the same amount of effort, and their income increases commensurately. See Figure 1 for a comparative analysis of time invested and results.

Figure 1: Broker vs. Salesperson Time Invested and Results

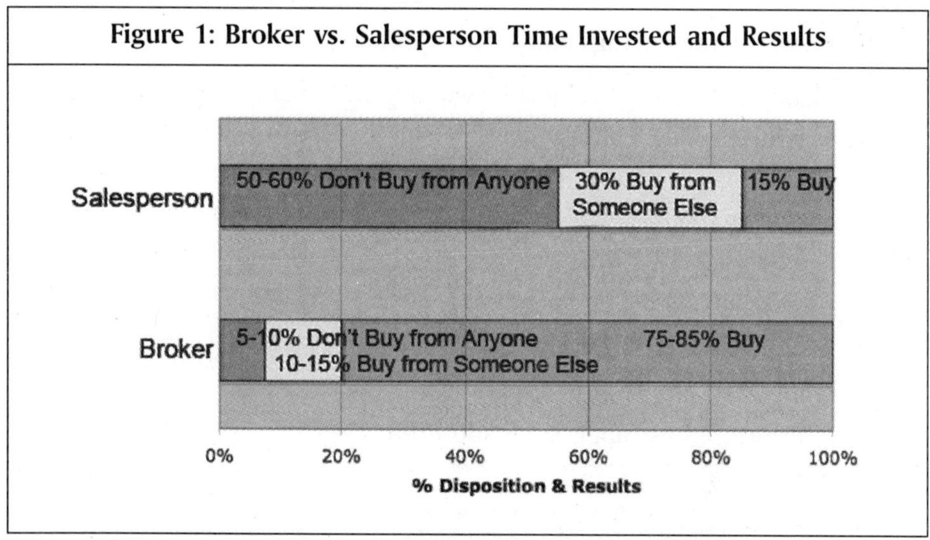

It should be noted that the installer's time saved might not be quite 100% because they still need to prepare a bid for the broker. And because they are going to win only 1 in 3 bids on average, they need to budget 1 to 1.5 hours per sale of time to prepare and email the bids. However, this might actually be a time savings, because they only need to prepare a simple quote (20-30 minutes each to prepare a quote based on the detailed specification and photos the broker provides), and rely on the broker to fairly represent them, rather than preparing full-blown proposals that might take a lot more time. See Figure 2 for an illustration of relative costs to the installer.

Figure 2: Seller Time and Expense Invested Sales & Marketing

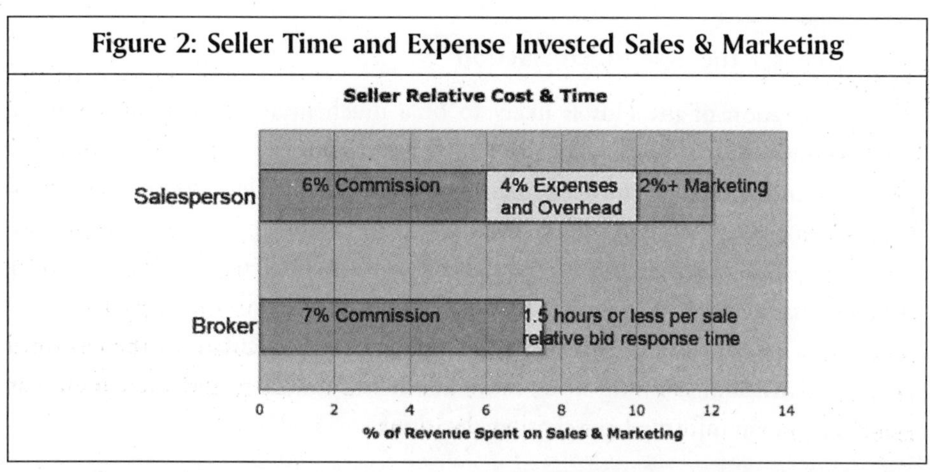

3.3 Purchaser Time Savings and Effort Reduction

The purchaser saves significant time by working with a broker, because they now only have to meet with one solar professional for a site survey and bid presentation. They also only have to answer one set of customer interview questions during the phone screen. See Figure 3 for an illustration of the relative time saved for the customer. Because the broker is indifferent to the customer's ultimate choice, the customer may chose to confide more in the broker about concerns and motivations they might otherwise keep hidden from more biased salespeople. This can reduce the customer's need to spend time seeking unbiased information elsewhere.

Figure 3: Customer Time and Effort Invested in Purchase

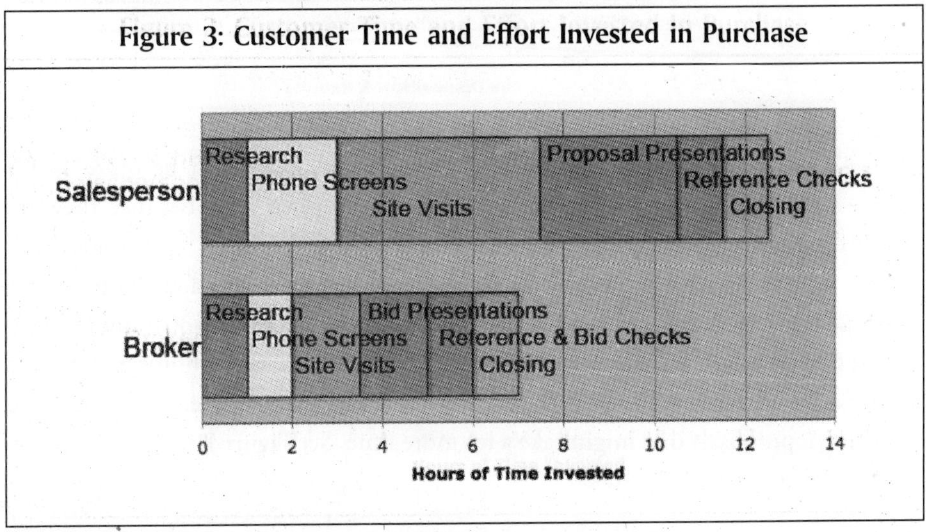

3.4 Purchaser Increase in Satisfaction

The presentation of the bids is likely to be a much more pleasant situation for the customer. Rather than being presented with 3 to 5 "pitches" attempting to persuade and influence the customer towards each installer, the customer and broker can have a low stress, low-pressure conversation, presenting all the bids side by side. The bids should be normalized, so that differences in wattage rating systems, assumptions for permit fees, sales tax, service entrance upgrades, and other costs are visible and made equal. These are often calculations the customer cannot do without learning a lot more about the industry, and even then, can result in poorly informed purchasing decisions.

Once the bids are normalized, and the customer and broker have had a full discussion about all the relevant issues, motivations and desires, the best choices become clear, and it is often just a choice between a couple of good options, rather than a nervous decision with a background fear of some unknown.

Also because the customer may be more likely to confide in the broker, the broker may have a better understanding of what the customer's true motivations, interests and fears are, and may be better able to provide the most satisfying solution by ensuring that the installer best equipped to handle a particular concern is chosen, leading to a satisfying and happy result for all three parties.

The broker can also serve as a mediator and problem solver, ensuring good communication and solutions flow between customer and installer. When a problem occurs, the experienced broker will be able to explain to the customer what's normal and reasonable, and what they can expect a good installer to do to remedy the situation, if it is truly the responsibility of the installer to handle the issue. If it's not the installer's problem, the broker can help explain the situation, and if necessary, defend and protect the installer from unreasonable customer requests.

The broker can be a powerful advocate for the customer. One can imagine an unforeseeable situation where things go poorly on a project and the cost of doing a good job may cause a financial loss to the installer. While most installers will accept the responsibility and the loss, some may take the approach that solving the problem properly just isn't worth the cost, and it's only one customer who isn't likely to buy again until they move in 5 to 10 years – this is always a risk when dealing with contractors directly. However, the broker has the choice to bring more business to the contractor, or cut them off. The broker is, in effect, a repeat customer, which can motivate the contractor to ensure the job is done correctly, even if the cost is significant and might cause a loss on this particular job.

3.5 Installer Increase in Satisfaction and Margin Protection

A broker can be a big benefit to the installer as well. The broker is, in effect, a repeat customer, which is rare in the solar industry. The broker can understand the installer's needs and situation and explain these to the customer without appearing defensive. The broker can also help in pairing and installer's hidden qualities and personality or style with the appropriate customers, so that the parties get on well.

The broker can also be on the lookout for troublesome customers and help the installer avoid them. A good broker will have no trouble turning down this sort of bad business because they will usually be very busy with good customers. Even if they aren't very busy, the time and effort spent on a bad customer will far outweigh the benefit they could have gotten from looking for new good customers.

The broker can also help protect the profit margin for the installer at the time of sale. Some customers are very aggressive on price, not realizing that the solar industry generally does not enjoy high profit margins. The broker is in a better position to explain this situation to the customer. If the customer persists, and attempts to create a price war between bidders, the broker can adopt a policy prohibiting or limiting this activity. The author has adopted such a policy, which will allow the customer to, one time only, request the broker to ask a higher priced bidder to match or beat a lower bidder. The higher bidder can re-bid or stand fast. If the higher bidder matches or beats the lower bidder, the customer must accept this new bid or lose the broker's services, and by the broker's agreement with the installers, lose access to any of these companies. If the bid doesn't come back lower, the customer must chose from the bids as they stand, or lose service. This avoids the "race to the bottom" situation that some customers attempt.

So while the bidders must be competitive, and the customer is assured of getting a fair price based on this competition, the bidders don't have to worry about having to take a dive on price or profit just to get the sale, and instead, can enjoy a reasonable margin on their work.

4. Risks in Brokering and Risk Mitigation

While there are a number of benefits to using a solar broker, there are also a number of significant risks to both the customer and installer that must be understood. The biggest of these is the risk that an inexperienced broker will make a mistake and that it won't be found until it's too late to correct easily. A group of competing sales people will each develop their own analysis and estimates of what will satisfy the customer's needs. If one of these is significantly off from the others, an adequately informed and alert customer will likely spot it and dismiss the errant bid. If only the broker is providing the analysis and determining system size and location, it is left entirely to the customer to check their work or trust their judgment.

4.1 Increased Risk to the Purchaser

Broker mistakes in the analysis of a solar system can include misestimates of usage, mismeasurements of site conditions such as shading, tilt or orientation, miscalculations in production or the time-of-use value of the production, which could result in either a misestimation of the needed system size or benefit. This could result in the installation of a system that is too large or too small to meet the customer's needs or expectations.

An oversold system that is too small to provide the promised benefit will leave the customer unsatisfied and disappointed with the system's performance, and might tarnish the installer's reputation or the reputation of the solar industry. Too large a system could result in wasted investment money and un-enjoyed production value from the system due to limitations on net metering benefit.

Or in the worst case, it could result in a bad purchase decision, where the customer would or would not have purchased if they had had accurate information.

The final risk to the customer is that an unscrupulous broker could collude with the installers and inflate prices above the reasonable and customary pricing for systems or misrepresent the installers' qualifications and experience. Because the broker is bringing potentially all the bids, such price inflation or quality flaws may not be visible to the customer.

4.2 Increased Risk to the Installer

An inexperienced solar broker might also expose the installer to substantial risk due to errors in the site estimation or analysis of the installation feasibility of the project. Inaccurate assessment of the sturdiness of the roof, or misreading the electrical service entrance label can result in substantial costs before the job can be completed in compliance with building or electrical codes. Mismeasurement of the roof area available, or other problems can result in an unbuildable job at any price.

4.3 Broker Requirements to Minimize Risks

Experience, training, and certification can be paths to minimizing the risks a broker poses.

Before engaging in broker relationships with customers or installers, the broker can gain significant experience in all the necessary areas by working in a traditional role as a solar salesperson. During this type of work experience, they gain knowledge of site analysis, customer interviewing, shade and other tool usage, performance estimation, and many other factors that will help them avoid making rookie mistakes as an unsupervised broker. During this time, they will learn from the mistakes that get found by those overseeing their work, or through competition with other salespeople upon presentation of the bids. The author estimates a broker should have at least one year of full time sales experience before considering starting as a broker.

The prospective broker can get on-the-job training working as a salesperson in a supervised position, but can also take some of the many courses offered by the training institutions available to the solar industry. The Institute for Sustainable Power Quality[1] (ISPQ) has established a list of accredited solar training schools, which is available at: *http://www.irecusa.org/index.php?id=91*

A high quality solar salesperson or broker should have a good understanding of the issues the installer needs to face, and ideally will have assisted in installing several systems to gain hands-on experience.

Certification, such as NABCEP® Solar PV Installer Certification[2] or CoSEIA PV Certification[3], are marks of qualification that indicate an individual has met the requirements of education, training and experience, and passed an examination demonstrating knowledge of how to safely and reliably install a PV system. Such certification might normally be above and beyond what the regular salesperson might want or need when selling PV in a competitive environment. However, because of the greater risk in working with a broker, such certification offers a valuable additional measure of security.

4.4 Customer Strategies to Minimize Risks

Consumers can protect themselves but using only brokers with strong word-of-mouth networks and relying on referrals from satisfied customers who have lived with their systems for at least a year. Checking the broker and installer references is always a good idea.

The consumer may also want to call a couple of other installation companies to get ballpark phone estimates to ensure the broker is in the right neighborhood on size, system estimated production, and price. Often, the phonescreener/estimator can discuss and look at satellite photos of the customer's home and get a sense if shading is a likely consideration. This plus the estimated production can indicate whether the broker has included shading in the calculations, or if they have overlooked a crucial piece of information. In the author's six-year design and competitive sales experience, shading is the most commonly unanalyzed or misanalyzed factor in system design.

It is important that the purchaser be clear that, because the broker is generally paid on commission by the installer, the broker's legal fiduciary responsibility is usually to the installer, despite implications that might be made otherwise. One might ask, why doesn't the customer instead hire the broker directly as a "buyer's agent". While this is possible, and likely for larger commercial projects, it is unlikely to be feasible or even available for most residential projects. The main reason is that it is unlikely that the customer would be willing to spend anywhere near 5-7% on a commission to a broker even though the broker does have significant costs not directly linkable to a particular sale, and would need to charge a high hourly rate to make up the difference. So high would the rate likely be, that the customer would probably just try to do it themselves, and solicit several bids through the normal sales channel. However, this eliminates all the efficiencies and savings the broker concept brings. Even if a buyer's agent were available, it is unlikely that these direct costs would be made up for by a discount provided by the installer unless the buyer's agent had established relationships with every installer and all relationships had equal terms. The author knows of no cases where a viable "buyer's agent" business has ever been attempted. Perhaps as solar brokering gets established, this will be a natural evolution for some customers, as it is in the real estate industry.

5. The Author's Experience

The author has been engaged in broker sales in the San Jose, California area from March 2005 to April 2007. Because of his experience, NABCEP® Solar PV Installer Certification, and the previous 4 years of direct sales leading to a well-established reputation and good word of mouth from previous customers, he has enjoyed plenty of broker business opportunities. However, because of significant teaching activity and the development of a solar sales and design software product, while he has only engaged a relatively limited number of clients, he has enjoyed a high level of success with the broker concept. He has closed sales for more than 75% of those who sought bids. Because the idea is a new business and approach, he believes that with refinement, the results can be even better.

6. Conclusion

While the use of brokers in the solar industry poses some risk, this can be mitigated by training and experience, and, the gains in efficiency for the industry, and reduction in costs for both the customer and industry, are likely to be significant.

The Solar Energy Industry Association's Roadmap[4] estimates that the solar industry reduces cost at least 4.5% per year (based on 35% compound annual growth in production volume which has occurred approximately annually over the last 7 years). If the broker concept can reduce industry selling price by 3.5% it's as if the whole industry jumped ahead by 6 to 9 months in terms of development down its experience-cost curve.

(Andy Black is CEO of OnGrid Solar, providing solar financial analysis tools and consultation. He serves as a board member of the American Solar Energy Society and the Advisory Board of NorCal Solar. Contact him at 408.428.0808 or www.ongrid.net. Black regularly teaches and consults on "Payback on Solar Electricity" for many audiences. The author can be reached at andy@ongrid.net).

7. References

(1) IREC – The Interstate Renewable Energy Council's website listing for Institute for Sustainable Power Quality (ISPQ) *http://www.irecusa.org/index.php?id=91*, April 2007

(2) NABCEP: The North American Board of Certified Energy Practitioners, requirements for PV certification, *http://www.nabcep.org/pv_installer.cfm*, April 2006

(3) CoSEIA: Colorado Solar Energy Industries Association, requirements for certification, *http://www.coseia.org/Certification.htm*, April 2006

(4) Solar Energy Industry Association Roadmap: Our Solar Power Future, p7-9, 12 *http://www.seia.org/roadmap.pdf*, September 2004

5

Solar Energy: Gearing up to Cope

Nirmala Rao Khadpekar

Solar energy which is known to be a clean and competent technology is over time becoming eminently possible to capture, in that the costs have been steadily falling. In the meanwhile, the technology is also being considerably refined in all the areas of operation. With the payback situations improving, solar is no longer considered a boutique option for those that can afford the indulgences in renewable sources of energy. While passive solar has been in use since millennia, the progress in technologies from simple flat plate to future orbital technologies, current solar energy has come a long way since the path breaking impact it made in mid-19th century. Social and environmental costs need to be seen against the long-term irreversible costs of the damage done by fossil fuels. Various issues related to these facts are examined in this article.

Introduction

When one thinks about it dispassionately, the possibility of managing an entirely clean energy production is low. If it produces no pollutant (emission) in its production, it is likely to leave a spot of waste[1] behind in its production or can

© *The Icfai University Press. All rights reserved.*

degrade the land though not grossly like what happens in fossil fuel capture. In short, since producing energy is a consumptive activity, in its excavation, drilling, capturing or transmission, it is so in its supply and use as well. There is something being supplied – unseen elements like air or gas – definitely not seen like hydro/tidal, mined coal and whatever potential lies in the subterranean rocks. Energy also costs to extract, it has to use other forms of energy/equipment/material for its capture and storage. It costs in terms of investment in technology, preceded of course by R&D, and then for equipment, installation, operation, maintenance, storage, transmission, with the need for a back up technology, requiring more energy and availability of trained manpower. The trick is in planning for it intelligently, and delinking it from odious comparisons with covertly subsidized fossil fuels which we have been enjoying all this while at great cost of the environment. Over time, the perception of renewables has changed. Solar power is no longer considered a boutique technology but as practical, affordable, and reliable; accessible by all from the deep pocketed and 'aware' person who can afford to indulge in redressing climate change to the practical developmentally oriented and those in dire need of electricity supply in the remote rural. Solar has travelled the distance from the elitist margins to the mainstream. There is a clear benefit to the expansion of solar technology development and implementation as it has the advantage of being a proven technology which is at refinement stage. By 2020 installed solar capacity is likely to be 20 to 40 times its present level[2].

'Shock and Trance' Routine

The interest in alternative energy or renewable resources peaks every once in a while, especially when the oil prices soar. At the time of planning this book, oil prices were soaring, but at the time of writing this article, there has been an economic downturn and oil prices are falling. It is hoped that the recently renewed interest in clean energy generation and supply do not wane again. As declared in mid-November 2008 by no less than the president elect of North America, Mr. Barrack Obama, doing something about is more important now but may be just a little harder politically, "...this has been our pattern. We go from shock to trance. You know, oil prices go up, gas prices at the pump go up, and everybody goes into a flurry of activity. And the prices go down.....and we start filling up our SUVs again......as a consequence we never make any progress.....it's part of the addiction.....now is the time to break it[3]." Nobody owns the sun or wind, which is considered a by product of the sun. Both have been up there in the atmosphere since the beginning of time and have fascinated mankind for millennia. There is worship also involved here. In this article we look at the sun. In the 5th century BC[4] the Romans used passive solar energy in their architecture. In 1760, the first prototype of the solar cooker came in and in 1874, a French engineer used steam produced from a solar boiler to operate a power engine to pump water in Algiers[5]. In the 1890s solar heated pools were developed in the US[6]. Despite this history, solar power generation is really a mid 19th century technology which was first taken forward by Bell Labs[7]. But it was not till 1970,[8] when the price decline began that matters began looking up for the sector. Of course, the boost from subsidy and push from investment and tax credit plans are required. Rich quality savings will accrue in reduction of greenhouse gases for developed countries, and improvement in quality of life for developing countries who cannot reach power to the distantly located populace. In countries where the supply of electricity is linked to the base livelihood, the concerns are more economic than environmental as has been recognized by international protocols and world funding agencies.

Social Costs

The social cost of climate change in addition to the economic fallout due to environmental degradation is huge. The social cost of carbon is the damage done by emitting an additional unit of carbon dioxide which is expressed in US dollar per metric ton of carbon. According to Prof. Richard S J Tol, "The social cost of carbon is thus a measure of the seriousness of climate change, and a yardstick against which to judge actual and proposed climate policy."[9] He further works it out like this:

- future development of population and economy;
- future emissions of greenhouse gas emissions;
- future climate change;
- future impacts of climate change;
- values of these impacts;
- the rate of pure time preference;

- the rate of risk aversion; and
- the rate of inequity aversion[10].

The variables are defined as 'positive variables placed in the future, a mix of positive and normative elements, which can be controversial as well as uncertain.'[11]

Prof. Tol has conducted a study titled the Social Cost of Carbon Outliers and Catastrophes[12] in which he discusses that how there is a 'fair chance that the annual climate liability can exceed the annual income of many people.' He has presented 211 estimates of the social cost of carbon in a meta analysis where he discusses the results how lower discount rates could imply a higher estimate. And how there could be a descending trend in the economic effect on the approximations of the climate, quoting from the Stern Review's statements about the social cost of carbon.

Practical Benefits

Once put in place, the solar system has near nil variable costs much like that of hydroelectric or nuclear power. It is also not effected by price volatility which means that it skirts commodity price movements as also carrying no forward exchange risk during its life[13]. Solar energy is also credited with the highest reliability factors within the alternative options. It can step in during grid downtime, and can also access premium power markets. Practically also it enjoys a good match with the public utility load requirement as it corresponds well with summer electricity high load. It is also amenable to distributed generation in which method the amount of energy lost in transmitting electricity is reduced because the electricity is generated very near to where it is used, possibly in the same building itself. The size and number of power lines required to be constructed also comes down considerably. The favoured location again is the rooftop of buildings.

The impact on the economy is also a double benefit as it is a driver spawning a whole new industrial set up to supply the new technology requirements. The equipment industry as portions of the solar construction industry is considered labour intensive and generally involves materials development within the region, making it a significant job creator especially for installation and maintenance activities for the local people[14]. Specialized expertise in solar engineering, planning, and architecture are the dynamic result – a new specialization emerging as it were. This could also happen locally with adequate planning as these set ups too need a point to point production (20 MW per year) to attain the efficiency curve to be profitable. Further there is always ample rooftop space generally unutilized for anything else for spreading out upon as well. However, the concentration needs to be now on market enlargement to tap the economies of scale to continue the price decline[15].

The people who find installation of a solar electric system appealing have been generally seen to be having consumption bills of over $100 (approx Rs.5,000 per month) and therefore find the cost of installation much easier to justify. Michael Bishop, in his paper, "A Solar Energy System in your Portfolio: The Residential Case" categorically declares that the purchase of the solar electric system is one of the most secure investments people can make. He also says that 25 year warranties are now commonplace – which is an unprecedented commitment in the technology industry. He lists the variables and the special physical features including shading issues in his paper along with how property resale values are enhanced when solar installations are built in. The McKinsey Global Institute (MGI) has published a comprehensive study on the economics of investing in energy productivity. In essence, it prompts public and private sectors to lead the way by 'setting efficiency standards for appliances, equipment, financing energy upgrades, raising corporate standards for energy efficiency, and collaborating with energy intermediaries.'

Sound Financial Side

Solar energy is applicable in agriculture, industrial, residential, urban planning and motor vehicles. The economics of the sector is changing through solar energy capital and generation cost is currently high, three to four times more than the cost of conventional energy. (See Figures 1 and 2) Costs in this sector have been declining on an average of 4 per cent per annum and gaining acceptance worldwide as renewable technologies improve with economies of scale – from cookers to satellite technology. In many circumstances, when the user option of connecting to an electricity grid is absent, solar is the energy of choice[16]. This market segment is providing the economic platform from which a self-sustaining commercially driven, high technology industry is emerging[17]. There are sound economic principles at work in the math. Continuous advances in PV technology and economies of scale in manufacture have been making the investment more and more realistic and reducing the prices steadily. Breakeven costs with other grid supplied power are becoming feasible with time and some predict breakeven will occur by the end of this decade. There is potential for solar to become a mainstream option in the foreseeable future according to many researchers. The economic backdrop is so promising that even conventional financial consultants have begun to see investment in solar energy as an 'intelligent addition to a well-rounded investment portfolio'.

Figure 1: Cost of Different Renewable Forms of Energy (Global Average)

Technology	Typical Characteristics	Typical Energy Costs (US cents/kilowatt-hour)
Power Generation		
Small Hydro	Plant size: 1-10 MW	4-7
On-shore wind	Turbine size: 1-3MW	5-8
Off-shore wind	Turbine size: 1.5-5 MW	8-12
Biomass Power	Plant size: 1-20 MW	5-12
Geothermal Power	Plant size: 1-100 MW	4-7
Rooftop solar PV	Peak capacity: 2-5 kilowatts-peak	20-80
Concentrating Solar Thermal Power (CSP)	Peak capacity: 50-500 MV (trough), 10-20 MW (tower); Types: trough, tower, dish	12-18
Rural (off grid) Energy		
Mini- Hydro	Plant capacity: 100-1,000 kW	5-10
Micro-Hydro	Plant capacity: 1-100 kW	7-20
Biomass gasifier	Size: 20-5,000 kW	8-12
Small wind turbine	Turbine size: 3-100 kW	15-25
Household wind turbine	Turbine size: 0.1-3 kW	15-35
Solar home system	System size: 20-100 watts	40-60

Note: Costs are economic cost, exclusive of subsidies or policy incentives.

Source: REN21, Renewables 2007, Global Status Report.

Figure 2: Energy Market Share & Life Cycles 1860-2060

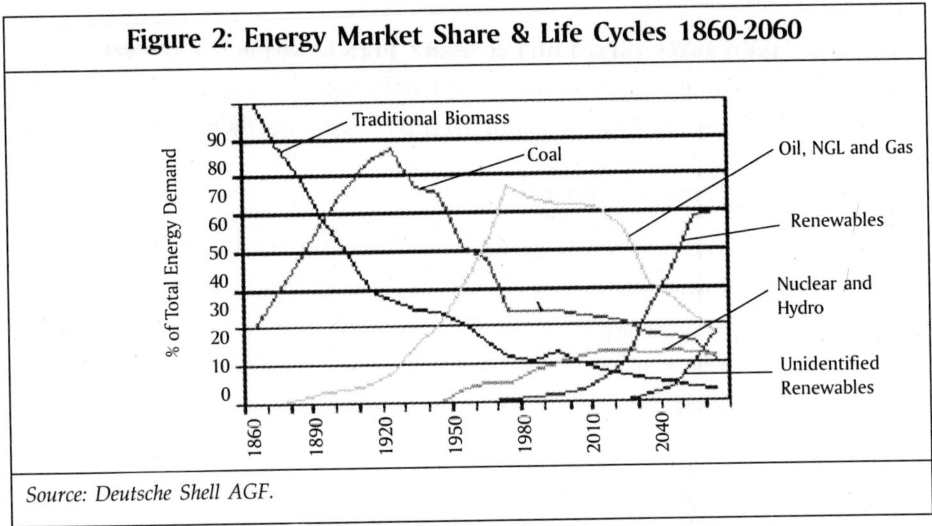

Source: Deutsche Shell AGF.

Well known venture capitalist Vinod Khosla, who is best known for founding Sun Microsystems, one of alternative technology's most ardent investors, is a known supporter of solar energy. "He has funded entrepreneurs building solar power plants that will dwarf football fields and companies that will make ethanol from wood chips," says an interview with him in *On the Record*, May 11, 2008. The discussion is also about the future of green technology and which technologies will thrive and which will die out. Silicon Valley may be concentrating more on solar than computing in the days to come. Cheaper alternatives to fossil fuels need to be innovated, developed; then simple mathematics will help their proliferation all over the world. "The world's venture capitalists, having been fed on the computing boom in the 80s, the internet boom in the 90s and the nanotech and biotech boomlets in early 2000 are now looking for the next one. They think they have found it: energy," says *The Economist* in 'The Power and the Glory' June 2008. They also add that the reasons for the boom are complex and the way they are professed may change, as global warming is after all a long range occurrence and may not be on the top of the public mind in economic downturn time. Renewables as investments cannot be simple and quick course corrections within the manufacturing process. The probability needs a whole new investment and tooling for a shift to a different technology all together. The Stern Report[18] author, British economist, who is a former World Bank chief and author of the 2006 Stern Review, has advocated on a visit to India in December 2008, that Tata's Nano car should be engineered for biofuels. Now that is not only a macro order for a micro car, it means engineering it all over again, even if it is an excellent idea.

Technology Development

Technology development and research in the solar area have concentrated on finding optimal methods to capture and develop system to convert it into usable applications. The power areas of significance are wind, biomass, tidal and hydro, most of which have not been adequately tapped till date. A total of five significant technologies utilizing solar are under development. The point of concentration is improvement and making the technologies more efficient presently in the areas of silicon-based wafer thin filmed photovoltaic, and concentrated solar thermal. To elaborate:

- Heat for buildings, for electricity generation, and industrial processes.
- Photovoltaic cells to convert solar energy into electricity directly.
- Exploitation of chemical reaction produced via photosynthesis triggered by solar radiation to provide heat or multiple uses.
- Mechanical energy from wind energy systems to be converted into electric power.
- Ocean thermal energy conversion which utilizes the difference between warm and cold levels of water to generate electricity.

Ingenious Storage Systems[19] become an efficient way of segmenting the direct use and supply of energy as weather can have an impact, and continuous night time supply can also be assured. The four storage methods are:

- Storing of thermal energy in materials with high heat retention properties called sensible-heat-storage-systems
- Storing solar energy in vaporization of (latent heat fusion) of particular materials experiencing a change in phase
- Chemical storage of energy using reversible reactions
- Use of storage batteries either electrical or compressed air.

Photovoltaic systems have the advantage of being modularly designed and have an inherent versatility. We are familiar with small energy in wristwatches, calculators powered by solar. This potential has now been extended to entire communities, power plant generators, in dispersed applications and of course, for remote settlements unable to be serviced by connections to electricity grids. New technology breakthroughs as reported by 'The Guardian' in 'Solar energy brings green power closer' December 29, 2007, shows how a solar plant of 10 MW size could be put up and how more such economies of scale could become more and more attractive. Crystalline silicon which has been the material of choice for PV has been taken forward to thin mono-crystalline cells called *Sliver Solar Cells* which consume only 5 to 10 per cent of pure silicon[20]. These thin modules are meant for low cost, bifacial, transparent, flexible shadow tolerance and lightweight applications[21]. Researchers at Massachusetts Institute of Technology have reported promising results for new technology that could make it easier to store energy, by enabling cells to continue to generate electricity under cloud cover at night, according to *McKinsey Quarterly*, August 2, 2008 'Economics of Solar Power'.

New, economical methods to bring in fresh solar energy products into the marketplace that are more reasonably priced and simpler to set up than those accessible today need to be developed. Tokyo Ohka Kogyo Co Ltd (TOK) and technologists at IBM's New York laboratory, Yorktown Heights, are working together to jointly develop new processes, materials, and equipment suitable for the production of CIGS (copper-indium-gallium-selenide) solar cell modules. With the cost of solar produced electricity still higher as compared to traditional fossil fuels, the barrier to wider adoption still exists. Use of thin film technology (CIGS) in the production of solar cells and elements holds hope for reducing the costs on the whole. "IBM has reported that its researchers have developed non-vacuum, solution-based manufacturing processes for CIGS solar cells with targeted efficiencies of approximately 15% and higher, compared to current thin film product efficiencies that vary from approximately 6% to less than 12%."[22]

Statutory Matters

A few countries across the world are also establishing grounds to mandate use of PV in new construction through building code regulations. The year 2006 saw such activity in Spain which included renovated buildings. In California, new initiatives require construction companies to offer solar installations as a standard feature in developments of 50 plus unit in townships/homesteads. Governmental subsidies and grants have a major role to play in kick starting the leap into such decisions; California, Italy, Spain, Germany are some of the developed markets. The market is still evolving and it will be prudent for utilities and regulatory bodies to make intelligent choices to enable nurturing of this opportunity thrown up by the fact that traditional energy sources are becoming more expensive, toxic and politically unviable as well. The importance of requirement for regulation makes some feel that solar could perhaps be technologically uncertain, but on the whole its economic attractiveness is improving steadily. In terms of utilities, electricity producers and the equipment manufacturing sectors, governments and regulatory authorities have a large responsibility. Encouragement of use of electric and hybrid vehicles by governments will be another step forward in prioritizing a basket of goals. Production rewards and subsidy phase outs have to be planned for a maximization of prolonged benefits for the sectors and users/installers. China and India[23] are expected to set the pace for the sector's growth. India has also gone a long way in using solar energy. A case in point is the widespread use of community solar cookers for indoor cooking, which allows high volume solar cooking in a convenient, shaded environment. The world's largest Solar Cooking System is located at Tirupathi Temple. It has been installed on the terrace of Nitya Annadanam Canteen, and offers a unique solution that takes both ecology and economy into consideration. This shows that along with high technology, developing countries need to look into practical propositions to make areas like mass/poor feeding campaigns feasible on larger scales.

From Outer Space

An energy that is free, abundant and inexhaustible, capturing it has no negative environmental fallouts. The future lies in focusing on solar engineering to improve technologies to harness solar energies and improve the applications. Creation of economies of scale is the magic answer to make such technology viable. The enlarging population with ever increasing consumptive needs of power dependence; all have a roof under the sun, the great giver of life to a cold planet. Now this planet is reeling under the horrors of climate change; with the very same sun pointing a path making it lucid for technologists to illuminate their minds with innovative solutions. The sun belongs to everyone; it has no political airspace, unless the impending technology of farming energy from space brings in a few conflicts and dilemmas. There are reports about solar research in Japan as mentioned in the article 'Farming Solar Energy in Space'. Despite massive costs, Japan pursues solar energy from space program says Tim Hoyak, July 1, 2008. Researchers at the institute of Laser Technology in Osaka are working on projects that could produce up to 180 watts of electricity from sunlight. Scientists in Hokkaido have also begun ground tests of power transmission systems to transmit energy from the Sun to Earth in microwave. The laser and microwave research projects are two parts of a bold plan for a space power system under the umbrella of Japan's space agency which aims to put into geostationary orbit a solar power generator that will transmit one gigawatt of energy to the earth, equivalent of the output of a large nuclear plant. Now what should be construed from that and how will the oil lobbies react? In the meanwhile, there are plenty of terawatts to be organized and a lot of money to be invested and to be made, on the cards. The efforts are on in making user friendly and easy top install technology for dish capture of solar energy collection[24]. The environment is dynamic as long held beliefs about technologies and subsidies are being re-examined. The efforts are to attain true grid parity for solar and other renewables, when the price match takes place.

(Nirmala Rao Khadpekar, Senior Faculty Member, Icfai Research Centre, Ahmedabad. The author can be reached at nirmala.khadpekar@gmail.com).

Endnotes

1 *www.ChinaPost.com.tw*

2 "The Economics of Solar Power", Peter Lorenz, Dickon Pinner, and Thomas Seitz, *The McKinsey Quarterly*, June 2008.

3 "Dot Earth", *The New York Times*, November 17, 2008, interview with Andrew Revkin on 60 minutes.

4 *http://southface.org*

5 Solar products at Ecomall.

6 *http://californiasolarcentre.org*

7 *http://www.eeere.energy.gov*

8 Chapter 10, Solar Energy, The Energy Report, May 2008, The Texas Comptroller of Accounts.

9 *Topics.scirus.com*, last updated September 03, 2008.

10 Ibid.

11 Ibid.

12 *www.economics-ejournal.org/economics/journalarticles/2008-25/view-8k-*

13 *www.solarbuzz.com*

14 *www.azsolarcenter.com/*

15 *seekingalpha.com/article/55571-first-solar-s-rapidly-expanding-economies-of-scale*

16 *www.solarserver.de*

17 *www.solarbuzz.com*

18 *http://www.peakoil.com/article37516.html*

19 See "Molten Salt may be Solution to Solar Energy Storage" by Carol Gulyas, June 29, 2008 in CleanTechnica.

20 "Sliver Solar Cells", *Proc. Of SPE* Vol. 6800 10-3.

21 Ibid.

22 *http://www.allbusiness.com/electronics/computer-electronics-manufacturing/*

23 India's National Plan.

24 The Good Human 2006-2008.

Section II

Technology Alternatives

6

Technology Advances in Delivering Cost-Competitive Solar Energy

The prevalent approach for generating solar electricity is through photovoltaic (PV) systems that use semiconductor PV material to convert sunlight directly into electricity. The market is currently dominated by silicon-based flat-plate PV, which produces electricity by having sunlight directly strike panels tiled with sheets or wafers of expensive PV cell material. Flatplate PV has defined the solar electricity market since its emergence 30 years ago, and drove the market's recent rapid growth to $7 billion in 2005. However, recent silicon feedstock shortages and rising wafer prices have highlighted the need for technologies which can advance the market beyond current flat-plate PV and its underlying cost limitations.

Concentrator photovoltaic (CPV) approaches enable lower total system cost by reducing the amount of PV material used. They use mirrors or lenses to focus sunlight collected from a given receiver area onto a much smaller area of PV material, and in some cases use higher efficiency, non-silicon PV material. This paper describes a particular high-performance CPV solution which compared to average flat-plate PV, uses

Source: www.parc.com © 2008 PARC Incorporated. Reprinted with permission.

1/500th as much PV material and produces nearly twice as much electricity for a given collection area. The proposed solution is also smaller, cheaper, and easier to manufacture. These improvements enable generating electricity at less than half the cost possible with existing flat-plate technologies, potentially opening large new markets for clean solar energy.

The Market for Solar Electricity

Rising fossil-fuel prices, recent geopolitical developments, and environmental concerns have led to growing demand for renewable energy sources such as solar. Not only is solar energy secure, widely available, and carbon-free, but more energy from the sun strikes the earth every hour than is consumed on the planet in a year [1].

Though sunlight is considered a "compelling solution" to the "need for clean, abundant sources of energy," solar energy currently provides only 0.01 percent of the total electricity supply [1]. However, recent market trends, regulatory pressures, consumer incentives, and technological advancements are together driving solar energy costs down relative to conventional fossil fuel-derived energy.

Increasing global investor interest has helped bring solar energy into the mainstream. In 2005, solar investments were thirty times higher than a decade ago and already twice the 2004 level. Solar sector investments represented more than one-third of the total money invested by venture capital firms in the entire US energy industry [2]. Furthermore, the top three initial public offerings (IPOs) of 2005 were all solar companies. As one solar entrepreneur put it, solar is the "next big disruption" and "dwarfs any business opportunity in history" [3].

Bold subsidy programs and regulations—including feed-in tariff laws, renewable portfolio standards, tax credits, and rebates—in a few key markets such as Japan, Germany, and California have significantly contributed to lowering consumers' solar costs over the past 30 years. In Japan, subsidies have gradually been reduced and solar electricity is considered to be competitive with conventional electricity [4]. The Japanese market is expected to maintain solar growth and low prices without subsidy, and the burden of maintaining low costs has now passed onto manufacturers [5].

While Japan has the most installed solar electricity, Germany is the fastest-growing market. But California's expansion of its solar incentive program by $3 billion in January 2006 has made it one of the most aggressive in the world, and a number of US states are also expanding or considering their solar and other renewable energy options. China also initiated a solar subsidy program in early 2006, with the goal of making solar energy a critical component for addressing the exploding energy needs of the significant Chinese market. Substantial new programs are also under way or being considered in Australia, India, Portugal, South Korea, Spain, and other large markets.

Compared to conventional and other renewable energy sources, solar power is especially attractive because it can be easily scaled down or up. Solar electricity can also be generated nearer to consumers and even on site, which greatly reduces or eliminates transmission costs. Furthermore, the increasing adoption of variable-pricing or net metering schemes also favors solar electricity. Under these schemes electricity rates are higher when peak demand is highest, and this generally correlates to when more solar energy is available and electric output highest.

Solar costs are also presently being lowered through higher volume production, improved manufacturing techniques, and alternative solar technologies that reduce the amount of semiconductor material. Total installed system costs can further be reduced through cheaper "balance-of-system" components such as inverters and labor (which can be reduced through improved design and installation techniques).

However, there is consensus in the industry that broad market success for solar will come only when costs are competitive with conventional fossil fuel-derived electricity—without subsidies. The bold incentive programs and rapid solar market growth described above have already lowered the effective cost of solar PV systems to consumers. This in turn has driven the higher volumes that have lowered actual manufacturing costs, making solar electricity even more cost-competitive. Rising fossil-fuel prices have further reduced the gap between conventional and solar electricity. But most of the technological advances thus far have only been incremental, and without a dramatic technological advance like the solution proposed in this paper it would take another decade or more for solar costs to get within striking distance of conventional electricity.

Flat-Plate PV Dominates the Market

The prevalent approach to solar generation of electricity is through "photovoltaic" (PV) systems which convert sunlight ("photons") directly into electricity ("voltage")[1]. In fact, grid-connected PV is the fastest-growing energy technology in the world. In just a four-year period (2000-2004), capacity grew at an average of 60 percent a year [6].

Solar PV cells—the basic semiconductor device for converting sunlight into electricity—are usually grouped together and electrically configured into modules, panels, and arrays. The dominant technology for these cells is based on single-crystal or polycrystalline silicon. "Flatplate" PV panels produce electricity when sunlight directly strikes the surface area of a panel tiled with sheets or wafers of expensive silicon PV cell material. Flat-plate silicon PV has been the primary solar product on the market for the last three decades and during the market's recent rapid growth to $7 billion in 2005. Most installations today are on rooftops, but larger field-based installations of 10 megawatts (MW) or more are becoming common, and there are several plans for installations in the 100 MW range.

The silicon material cost accounts for a substantial portion of the total installed PV system costs, and is generally considered one of the best opportunities for lowering overall costs and increasing the competitiveness of solar electricity. Furthermore, given the size of the global energy market and the ultimate potential of solar as a conventional electricity source, even a moderate reduction in the cost of solar can enable further market expansion.

Despite widespread interest and support for solar PV systems, cost and performance advantages are therefore needed to advance the market beyond current flat-plate PV and to make PV competitive with electricity from conventional sources.

---

1 Besides PV where sunlight is used for direct electrical conversion, other solar power approaches include solar thermal energy where heat is used to drive turbines and generate electricity.

The Opportunity for Concentrator Photovoltaic (CPV)

Flat-plate silicon PV is dominant today, but silicon feedstock shortages and rising wafer prices have highlighted the need for technologies which can lower costs by reducing the amount of silicon or other semiconductor PV materials. To significantly change the economics of solar electricity requires alternatives such as thin-film solar technologies or concentrating technologies [7].

Thin-film solar technologies use semiconductor materials such as CIGS (copper indium gallium selenide), a-Si (amorphous silicon), or CdTe (cadmium telluride) and are fabricated with cheap roll-to-roll processes. Though these non-silicon technologies have received a fair amount of media and research attention, they face a number of financial and technical limitations including more use of scarce and expensive source materials (e.g., gallium and indium), modest efficiencies, and reliability issues. Concentrator PV (CPV) approaches, on the other hand, offer an effective, practical way to keep solar cell conversion efficiencies high while keeping semiconductor material costs down.

CPV technologies use relatively inexpensive optics such as mirrors or lenses to "concentrate" or focus light from a relatively broad collection area onto a much smaller area of active semiconductor PV cell material[2]. Since the PV semiconductor material dominates the costs of the solar PV system, reducing the amount of PV material required to capture a given amount of sunlight leads to substantially lower system cost and resulting cost per watt of output [8].

Some CPV modules reduce the amount of PV material used by as little as 2-5 times and for these, silicon is typically used as the semiconductor material. For higher concentrations that reduce PV material use by 100-1000 times more, it can be cost-effective to use higher efficiency cells that increase the electricity generated from a given collection area—even though this PV material can cost up to ten times more than silicon. Currently, the most efficient approach is multi-junction cells that achieve up to 40 percent conversion of the sun's energy into electricity, by using stacked layers of III-V compound semiconductor materials

---

[2] All concentrator systems—including CPV, Stirling dish engine systems, and solar-thermal-electric turbine systems—operate on the same concentrating principle, but the given receiver area over which sunlight is gathered ranges from a few centimeters for CPV, to meters or tens of meters for Stirling systems, to hundreds of meters for other solar-thermal-electric systems.

to capture more of the solar spectrum. In addition to enabling more electricity from a given amount of sunlight, multi-junction cells obtain even higher efficiencies at higher concentration levels.

Since CPV systems work by focusing sunlight onto a targeted area, they must be pointed directly at the sun and require trackers that follow the sun's trajectory throughout the day. Though this has historically been an adoption barrier for CPV relative to flat-plate PV, there have been a number of cost and reliability improvements, and trackers are commonly used in large field-based installations of even flat-plate PV. CPV has generally been viewed as most suitable for utility-scale applications, where tracking systems would be mounted on land to generate 10-100 MW of electricity. However, improved system features including lightweight trackers and recent technological advancements are allowing CPV to be mounted on rooftops, where it can generate 10-100 kilowatts of electricity at many commercial sites or 1 MW or more on large industrial rooftops.

SolFocus: Breakthrough Concentrator Photovoltaic (CPV) Incubated at PARC

The Palo Alto Research Center (PARC) recently signed a broad agreement with California-based start-up SolFocus to jointly develop CPV systems that can deliver low-cost, reliable solar energy.

For over three decades PARC has been conducting interdisciplinary research in the physical, computational, and social sciences to deliver innovations with proven commercial value. PARC today is applying its unique combination of competencies to provide strategic research services, technology, and intellectual property to diverse industry and government partners. PARC is also pursuing broad initiatives in areas such as biomedical sciences and "clean technologies" or Clean Tech.

One particular thrust of PARC's Clean Tech initiative is reducing the cost of solar energy, and PARC scientists have been applying expertise in optical system design, optoelectronics, and advanced materials and processes for electronic packaging to address this problem. The SolFocus venture allows PARC to directly contribute to the solar energy industry, participate in broader markets, and continue developing high-impact technologies.

Building on its relationship with PARC, and with strong early contributions in optical design [10, 11] from researchers at the University of California at Merced, SolFocus has already developed two related, low-cost CPV module designs: Gen 1 and Gen 2. Up to 2 MW of the Gen 1 design will be installed in 2006-2007 at pilot sites in California, Hawaii, and Shanghai, China. The Gen 2 design further improves performance and reduces costs at higher volume production, and will be available for test installations a couple of years later. Both module designs are targeted for rooftop- and field-installed solar systems, and the Gen 1 design may also be used for other applications such as augmenting conventional lighting in buildings.

CPV is increasingly seen as a compelling approach to solar electricity, and venture funding for CPV technologies is on the rise. Among available CPV solutions the SolFocus design is a strong contender for short- and long-term cost advantages, which is the critical metric for market success. Partly in recognition of this, SolFocus won the grand prize at the 18th NREL Industry Growth Forum which was sponsored in November 2005 by the Department of Energy's National Renewable Energy Laboratory (NREL). SolFocus CEO Gary Conley was recognized at this event as the Clean Energy Entrepreneur of the Year (2005).

The Technology

SolFocus' CPV modules are enabled by the highest-efficiency solar cells available—triple-junction cells with efficiencies approaching 40 percent at the operating point of 500-sun concentration. For a given receiver area over which sunlight is collected, the SolFocus CPV modules use only 1/1000th of the expensive multi-junction solar cell material to convert the same amount of sunlight as other PV systems. The modules track the sunlight with a two-axis system and then concentrate it using innovative imaging and non-imaging optics. They also optimize reliability and low-cost manufacturing with practices borrowed from the high-volume semiconductor industry. The SolFocus CPV modules therefore halve the costs and operate at double the efficiency of average flat-plate PV.

Specifically, both of SolFocus' CPV module designs:

- use glass—which can easily meet 30-year life requirements;
- use dry, passive cooling—thus requiring no liquids or fans;

- have modules with no moving parts—which avoids mechanical failure in the module;
- are fully "enclosed"—so there are no exposed mirrors or open fire hazards;
- use mirrors or reflective elements—which allow purely reflective light entry and avoid the chromatic aberration of lens-based concentrators;
- use minimal components—because they have a number of "double-purposed" materials;
- have one-quarter the focal length of other systems—making them extremely compact; and
- integrate the manufacturing capabilities of existing vendors—thus enabling the rapid launch of products.

While the Gen 1 prototype employs discrete optical elements, tailored imaging, and 1 cm^2 cells, the Gen 2 design incorporates small reflective concentrator elements in a single, flat, molded glass tile with mirrors on both sides; employs a non-imaging optical design; and uses 1 mm^2 cells.

Specifically, the smaller, thinner, "one-piece" Gen 2 design:

- uses the flattest optics—similar to the geometry of Cassegrain optical telescopes;
- reduces processing steps—by fabricating the optical elements in a single glass pressing;
- can be assembled with high-throughput, automated technology;
- packs flat for low shipping costs and associated benefits of offshore manufacturing; and
- avoids seals, filters, coverglass, and hard-to-clean surfaces [12].

Together, these CPV module designs yield dramatic improvements in size, durability, and scalability. They enable significant cost reductions, a quicker return on investment, and operate at higher efficiencies than average flat-plate PV.

SolFocus' CPV modules promise to lower the costs of solar electricity to less than half what is available today, and position PV to be competitive with electricity from conventional sources.

About PARC's Clean Tech Efforts

The Palo Alto Research Center (PARC)—the birthplace of ubiquitous computing, the graphical user interface (GUI), the Ethernet, the first commercial mouse, laser printing, and many other industry-transforming innovations—conducts interdisciplinary research in the physical, computational, and social sciences.

PARC is now applying a unique combination of its competencies to drive innovations in the "clean technologies" or Clean Tech sector. The Clean Tech focus areas listed below: 1) require broad competencies and multi-disciplinary teams; 2) allow PARC to participate in dynamic new markets; and 3) support PARC researchers' desire to apply technologies that have positive environmental and social impact.

PARC Clean Tech currently includes:

- *Solar Energy Generation Solutions*—including concentrator photovoltaic (CPV) systems and novel manufacturing approaches for silicon solar cells;
- *Energy Distribution Solutions*—including sensor networks for the intelligent power grid;
- *Energy Conservation Solutions*—including simplified, fine-grained sensing and climate control in office buildings;
- *Clean Water Solutions*—including post-filtration bio-agent concentration, fresh- and waste-water analysis, purity monitoring, and bio-agent detection (through electrostatic traveling wave-based particle manipulation, ultraviolet or UV emitters, and optical detector technologies);
- *Air Quality Solutions*—including contaminant monitoring (through electrostatic traveling wave-based particle manipulation, ultraviolet or UV emitters, and optical detector technologies); and
- *Paper Reduction Solutions*—including a new type of re-writeable paper that minimizes paper use for transient applications such as cover sheets, e-mail printouts, and other temporary documents.

References

[1] Basic Research Needs for Solar Energy Utilization. Report from Basic Energy Sciences Workshop on Solar Energy Utilization; 2005 April 18-21; Bethesda, Maryland. Washington, D.C.: Office of Science, US Department of Energy; 2005. Available at: *http://www.er.doe.gov/bes/reports/files/SEU_rpt.pdf*

[2] National Venture Capital Association. Cited in Associated Press. Venture capitalists embrace solar energy. MSNBC "Green Machines"; 2005 December 28.

[3] Bill Gross. Quoted in Reiss, S. The dotcom king and rooftop solar revolution. *Wired* (13.07); 2005 July.

[4] Solar Industry Statistics & Fast Solar Energy Facts. Solarbuzz [website on Internet]. San Francisco, CA: Solarbuzz, Inc. Available at: *http://www.solarbuzz.com/*

[5] Jones, J. (Editor). Japan's PV market. Renewable Energy World; 2005 February 3. Earthscan/James & James: London, UK.

[6] REN21 Renewable Energy Policy Network. Renewables 2005 Global Status Report. Washington, D.C.: Worldwatch Institute; 2005. Available at: *http://www.ren21.net/globalstatusreport/RE2005_Global_Status_Report.pdf*

[7] Beaver, M. "Missing the point on solar energy." *EE Times* (response to editorial); 2006 January 2.

[8] National Center for Photovoltaics [website on Internet]. Colorado: National Renewable Energy Laboratory. Available at: *http://www.nrel.gov/ncpv/*

[9] Elrod, S.A. (Palo Alto Research Center, Palo Alto, CA). Solar Concentrator Technologies: Capabilities in California. Submitted to Bay Area Science and Innovation Consortium (BASIC); 2005 November.

[10] Winston, R. and Gordon, J. "Planar concentrators near the étendue limit". *Optics Letters* (30): 2617-2619; 2005.

[11] Feuermann, D.; Gordon, J.; Horne, S.; Conley, G.; Winston, R. "Realization of compact, passively-cooled, high-flux photovoltaic prototypes". In Winston, R. and Koshel, R.J., Editors, *Nonimaging Optics and Efficient Illumination Systems II* (59420Q). Proceedings of The International Society for Optical Engineering *(SPIE)*, Volume 5942; 2005 August 20.

[12] Horne, S.; Conley, G.; Fork, D.K.; Shrader, E.; Radhakrishnan, R.; Gordon, J. Second generation reflective concentrator. Submitted to IEEE 4th World Conference on Photovoltaic Energy Conversion; 2006 May 7-12; Waikoloa; HI.

Other Sources

Horne, S.; Conley, G.; Fork, D. K. Progress in the development of modular reflective concentrators for large scale deployment. Submitted to SPIE Optics and Photonics; 2006 August 13-17; San Diego, CA.

Jones, J. (Editor). A time to concentrate. Renewable Energy World; 2005 February 9. Earthscan/James & James: London, UK.

National Renewable Energy Laboratory [website on Internet]. National laboratory of the US Department of Energy (DOE) Office of Energy Efficiency and Renewable Energy. Available at: *http://www.nrel.gov/*

Official Energy Statistics. Energy Information Administration (EIA) [website on Internet]. Washington, D.C.: US Department of Energy. Available at: *http://www.eia.doe.gov/*

Poletti, T. "California's rebate program brightens solar picture". *San Jose Mercury News*; 2006 January 23.

Red Herring; 2005 December 27. VCs double bets on solar.

"What's new in concentrating PV?" National Center for Photovoltaics [website on Internet]. Colorado: National Renewable Energy Laboratory. Available at: *http://www.nrel.gov/ncpv/ http://www.nrel.gov/ncpv/new_in_cpv.html*

7

2008 Photovoltaic Trends: Innovative Thin Film Technology and Large-Scale Power Plants; Gigawatt Perspectives in the USA

Rolf Hug

The article focuses on development of commercial solar power plants in megawatt category, developments in plant erection and innovations in thin film technologies and products that are conquering the global markets. Larger PV power plants are being built in Europe, USA and also in Asia.

Large-scale solar plants for the generation of electricity in the megawatt category as well as new photovoltaic thin film technologies and products are conquering the world: in Europe, in the USA and also in Asia more and larger PV power plants are being built. Industrial manufacturers of solar cells and of solar module manufacturing systems right up to turn-key solar factories are growing in the global market. In particular, turn-key solutions for the manufacturing of thin film modules are in demand throughout the world. In the run-up to the Intersolar

---

Source: www.solarserver.de © Heindl Server GmbH, Tübingen, Germany. Reprinted with permission.

2008, the 4th PV Industry Forum invites companies, suppliers and service providers of the solar sector to participate in a two-day specialised conference in Munich. The focus points of the event will be commercial solar power plants with megawatt capacities, as well as developments in plant erection and innovations in thin film processing. Another conference of the colourful supporting programme of the first Intersolar to be held in Munich will look into the solar potential of North America under the title of "Solar Gigawatts for the USA". Solarserver, being the official media partner of both conferences, in the Solar Report 5/2008 focusses on current trends in the thin film photovoltaic market and spotlights the solar markets in North America.

International Progress of Thin Film Modules

Material-efficient thin film technology that was continuously refined over the past years and now is able to produce layers of less than one micrometer, is gaining in importance side by side with crystalline solar technology, in spite of the somewhat more favourable situation on the silicon market: thin film modules are expected to reach a market share of up to 20% in a few years. According to estimates by market researchers about 100 photovoltaic companies throughout the world are working on the production of thin film modules with various technologies and materials and find themselves in different stages of research and development right up to industrial series production.

Thin film photovoltaic power plant "Baar" (4,8 MW) of the company EPURON GmbH.

SunFab Technology Centre of Applied Materials.

Sources: EPURON GmbH; Applied Materials.

One noticeable trend is that in the past two years companies from the electronics industry (semi-conductors and plasma technology) have increasingly turned towards photovoltaics. In particular, the US plant developer Applied Material and the Swiss Oerlikon Group have developed to offer potential manufacturers of thin film modules turn-key systems and to take over the extensive planning processes and the establishment of manufacturing lines for future producers. The Japanese solar cell manufacturer Sharp also turned to the promising future of thin film technology. Sharp, being the manufacturer of LCD panels, can resort to its extensive know-how of surface coating of glass and apply this to the production of thin film cells in the expansion of the production capacity of his Japanese factory Katsuragi. The Japanese manufacturers SANYO and Kaneka are also planning to expand their thin film capacities.

Industrially Manufactured Thin Film Technology as a Next Step to Enhancing the Competitiveness of Solar Power

Even leading manufacturers of conventional solar cells on silicon basis are now scrambling for thin film, for example Q-Cells, the globally leading supplier of solar cells. Announcements by Applied Materials and other suppliers of turn-key plants regarding the possibility of a decrease in costs on a US dollar per watt peak encouraged a number of new participants to enter the market. Within only two years, for example, it is said that eight new solar factories have been built in Taiwan alone.

Applied Materials as well as Oerlikon Solar boast a number of international clients. And both manufacturers of PV solutions are rapidly expanding: for example, at the beginning of 2008 Applied Materials took over Baccini S.p.A., a leading manufacturer of fully automated metallisation and test systems for manufacturing crystalline silicon. "This acquisition is an important step towards our goal of becoming the leading plant equipper in the solar industry", said Mike Splinter, president and chief executive officer of Applied Materials. The company's target is not only to lower the manufacturing costs of thin film modules and modules of crystalline silicon, but also to reduce the amount of silicon used in grams-per-watt of energy generated and thus to contribute to an improved competitiveness of solar power in comparison to conventionally generated electricity.

New Dimensions in Photovoltaic Production: System of the Manufacturer Applied Materials for Plasma-Enhanced Chemical Vapour Deposition, PECVD, for the Manufacturing of Solar Cells

Source: Applied Materials.

Integrated Production Systems are the International Trend

Applied Materials has among its clients the Chinese XinAo Group, Brilliant 234 – a 100% subsidiary of the solar cell manufacturer Q-Cell – and "Malibu", a joint operation of Schüco and E.ON. With a fully integrated production plant of Applied Materials, Malibu strives to improve the cost-benefit ratio by integrating PV modules into the building facades. The investor Good Energies (Amsterdam) and Norsun AS (Oslo) also want to enter this market: by establishing Sunfilm AG they established a joint operation with its head offices in the Saxonian town of Großröhrsdorf. On a manufacturing line from Applied Materials Sunfilm will be manufacturing tandem thin film photovoltaic modules on glass carrier materials of 5.7 m2. On 15.05.2008 Sunfilm announced that it would be erecting a second production line with Applied Materials technology and would be commissioning this line in about a year. With this expansion Sunfilm wishes to achieve an overall capacity of 102 MW. According to Applied Materials, a total of eight manufacturing lines with an overall capacity of 350 MW are currently being built, half of which have already been installed. At the beginning of 2008, another production line contract with a volume of one gigawatt was concluded.

Applied Materials introduced its production line for thin film solar modules for the first time in September 2007 under the name of "Applied SunFab".

The substrates used are four times the size of the largest thin film carriers currently used. "The Applied SunFab Line is setting new industrial benchmarks that can be used by clients worldwide to increase production capacities of solar modules and at the same time to achieve the lowest manufacturing costs per watt," Applied Materials emphasised during its presentation on the SunFab during the 22nd Photovoltaic Conference and Exhibition in Milan. The production line is suitable for solar cells with single as well as tandem junctions and can annually produce a sufficient number of modules for an output of up to 75 MW.

Applied SunFab production line uses substrates that are four times the size of the modules currently used.

A glimpse of future production of Sunfilm AG

Sources: AMAT; Sunfilm.

On the basis of the advantages presented by the carriers of 5,7 m2 the companies XinAo and Sunfilm want to lower manufacturing as well as assembly costs. "The use of these large substrates will accelerate the development of cost-efficient solutions to generate clean, renewable energies," said Franz Janker, executive vice-president of Applied Materials upon signing the contract with the Chinese XinAo Group.

Oerlikon Solar is also rapidly expanding its global presence. With clients that are already commercially manufacturing and other companies that are finding themselves in ramp up, as well as with over 300,000 solar modules produced, Oerlikon Solar had a solid foundation on which to build in the rapidly growing market of thin film production systems. The company in April already announced a doubling of the production capacity of its plant in Trübbach (Switzerland).

In September 2007, Oerlikon Solar introduced its micromorph tandem cell in the market. This latest development stage offers a significantly higher degree of efficiency in comparison to its amorphous solar cell predecessors and would further expand the conversion rate in the double-digit range and would thus enhance technological advantages.

Oerlikon Solar TCO System for the Manufacturing of Thin Film Silicon Modules

Source: Oerlikon Solar.

"Micromorph tandem is an important element in the reduction of production costs of solar energy and in achieving grid parity in the short term. This will allow this environmentally friendly technology to become an economically feasible alternative to conventional energy production. The target is to achieve grid parity by 2010," the company stated.

ersol and SCHOTT Co-operating in the Field of Thin Film Cells, Inventux Starting up in Berlin and CMC Already Producing in Taiwan

ersol Thin Film GmbH (Erfurt), a subsidiary of ersol Solar Energy AG, is already producing together with Oerlikon plants and has delivered its first modules in January 2008 already. In addition, ersol entered into a cooperation agreement with SCHOTT Solar GmbH (Alzenau) to jointly further the development of micromorphous technology for thin film solar cells. Both companies intend bundling their resources in research and development in order to attain a marketable product more rapidly and thus to achieve a leading market position in micro-crystalline photovoltaics.

The Mainz-based technology group SCHOTT is working together with ersol in developing thin film technology and is investing approx. 60 million euro in its Jena plant to establish a manufacturing site for thin film solar modules of its subsidiary SCHOTT Solar GmbH.

SCHOTT Solar Small-scale Manufacturing for Thin Film Technology in Putzbrunn near Munich

Source: SCHOTT.

The company Inventuz Technologies AG that was established in spring 2007 has chosen its future production site and head offices to be located in Berlin. Inventux is planning to invest about 49 million euro in a production site for solar modules. From autumn 2008 annually about 275 000 solar modules with a total output of 33 MW are to be produced. In the manufacturing process the raw material silicon is to be applied as silan gas 100 times thinner than in conventional crystalline technology. The contractual partner for this manufacturing process is the Swiss company Oerlikon Solar.

CMC Magnetics Corp. is to be the first client of Oerlikon Solar stepping up its production in Asia. The 40 MW turn-key production line was transferred from Trübbach (Switzerland) to Taiwan over the past months, was then assembled and adjusted in order to commence its pilot operation on 17.04.2008. The final commencement of production is planned for the second half of 2008.

Successful German Manufacturers of Photovoltaic Production Equipment: Medium-sized Companies are becoming Global Players

Even German medium-sized companies are actively contributing when it comes to exploring new markets for PV production plants. For example, centrotherm photovoltaics AG (Blaubeuren near Ulm) and the plant manufacturer Roth & Rau AG (Hohenstein-Ernstthal) are developing from medium-sized companies to worldwide suppliers of production technology. Centrotherm photovoltaics AG, provider of technology and services for manufacturers of solar cells and solar silicon, established a subsidiary in Taiwan and is currently building a new thin film research centre with a pilot production site for so-called CIGS lines. CIGS (copper indium gallium selenide) is currently the thin film technology with the highest degree of efficiency. In research this is currently stated to be at 19.9%.

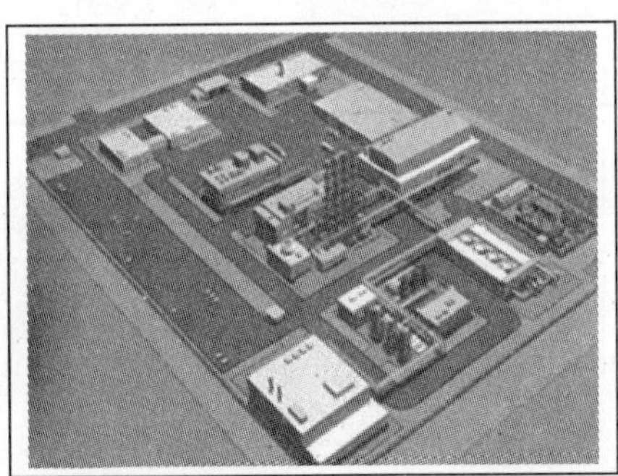

Model of a turn-key solar silicon factory of the company centrotherm photovoltaics

Solar cell coating system SiNA_L by Roth & Rau

The photovoltaic production site manufacturer Roth & Rau opened a new manufacturing site with additional capacities for its core product, the anti-reflex coating systems of the SiNA series. Furthermore, a manufacturing site for the production of thin film technology plants is being established. Currently Roth & Rau is the international leader in providing anti-reflex coating systems and fully automated.

Which of the new thin film technologies and production technologies will become established internationally not only depends on the quality and costs of the plants, a report of the Solarstrom magazine PHOTON emphasises. Because the development of thin film production could take a similar course as that of information technology. It is a known fact, that there the operating system with the fastest gain in market share won the race – and not necessarily the best operating system, Photon quotesGerd Lippold, the technology expert of City Solar AG. The worldwide race for thin film technology of the future remains open. The winner, however, has already been established.

First Solar at the Top of Thin Film Producers

The rapid ascent of First Solar – a manufacturer of thin film solar cells on the basis of cadmium telleride – is clear evidence of the ever-increasing importance of thin film technology. With a production of 200 MW in the year 2007, the company gained fifth position in the world ranking list of solar cell manufacturers. No other thin film producer was anywhere near the production figures of this US

First Solar Thin Film Modules

Solar Park Baar

Sources: First Solar (left), EPURON GmbH.

company that also has a factory in Frankfurt/Oder. The modules of the profitable thin film manufacturer are preferably used in large-scale plants of the megawatt category. For example, by COLEXON Energy AG, that is establishing the world's largest thin film roof system with an output of 4.6 MWp in Hassleben (Brandenburg); or in the future photovoltaic plant that is to be the largest of its kind, the energy park "Waldpolenz" (40 MW), in which the Mainz-based juwi group installed about 550,000 solar modules from First Solar that were mainly produced in Frankfurt/Oder. A total of 15 MW thin film modules from First Solar were installed by EPURON GmbH (Hamburg) during the course of last year. In Baar, EPURON connected another thin film park to the public grid in December 2007. The 4.8 MW project is among the Top Ten of the world's largest thin film plants. For EPURON it was the fifth solar plant that was realised in Germany in 2007.

Large-Scale Solar Power Plants in Europe, the USA and Asia

80% of all large-scale commercial photovoltaic power plants (regarding their output) are located in Europe which thus internationally holds the first position. Positions two and three are held by the USA and Asia. Within Europe, the German market has proven to be very dynamic and is thus one of the few that has shown continued growth during the past ten years. Currently almost half of all high-performance photovoltaic plants in the world are operated in Germany. Solarserver reported on these developments in January 2008 under the heading "Large-scale photovoltaic plants: 100% average growth since 2005". These and other developments in the photovoltaics sector will form part of the programme of the PV Industry Forum to be held this year. On the basis of the importance that photovoltaics now hold for German and international energy markets, the event was extended from one to two days. Also the increasing number of visitors and the increasingly international nature of this event required it to be extended in 2008: whereas in 2007, 319 decision-takers and experts in the field utilised the opportunity to exchange ideas and further their knowledge, more than 400 international visitors are expected this year. "The enthusiasm with which the field welcomed the conference over the past years has shown us that this platform for exchange is gaining in significance for the sector. With an extension of the PV Industry Forum we are hoping to meet this demand," Horst Dufner from Solar Promotion underlines, co-organiser of the PV Industry Forum.

"Solar Gigawatts for North America": Is the Northern American Market Ready for a Boom?

Only 5% of the world population live in North America. However, they use 26% of the entire energy produced worldwide for which they annually pay over 450 billion US$. But particularly the USA also have enormous potential of renewable energies, from the very windy plains of North Dakota to the Mojave Desert, an area that stretches from California to Arizona and is among the regions with the highest solar radiation in the world. Photovoltaics, solar thermal power plants, solar thermics, solar air-conditioning and solar construction in North America are the topics of the transatlantic symposium "Solar Gigawatts for North America". For the first time, this event, organised by Solar Promotion GmbH and the consulting company eclareon GmbH, informs about market developments and opportunities as well as about political framework conditions.

Open-space photovoltaics power plant with an output of 8.22 MW in Alamosa (Colorado)

Parabolic trough power plant "Nevada Solar One" in the Mojave Desert

Sources: First Solar (left), EPURON GmbH.

On 11 June the symposium will be discussing in Munich whether and how the ample available renewable energy sources can meet the demand and how the US market can develop in this direction, since the demand for renewable energy is rapidly growing. Currently, however, the continuation of tax credits that were granted to the end of 2008 is still at the centre of contentions. A new president will be elected in November. He could bring about a turn in energy affairs in the States and could awaken the "sleeping giant" solar energy in the USA.

Rethinking among US Energy Suppliers: Solar Power Gaining in Importance

Renewable energies, and in particular solar energy, are becoming increasingly popular in the USA and are now subsidised at almost all levels. Some states have developed own solar programmes, new companies have been established and are working on innovative technologies. The solar power and solar thermal market is taking shape.

Even large-scale energy suppliers – such as, for example, Southern California Edison (SCE) – are now looking towards photovoltaics. Other utilities are planning solar thermal power plants, for example, Pacific Gas and Electric. And this is done on an enormous scale: SCE alone is planning on investing about 550 m US$ in solar power plants. With the largest photovoltaics project of the USA approximately 6 million square metres of roof surface are to become solar power plants. Week by week SCE wants to install solar modules with an output of one megawatt. At this exemplary speed photovoltaic systems are to be installed on roofs of commercial buildings to achieve an overall output of 250 MW.

John E. Bryson, president and CEO of Edison International wants to build solar power systems on enormous roofs. This SCE plan resulted from the progress of photovoltaics that significantly reduced the installation costs of solar power plants, says Bryson.

Source: Southern California Edison.

SCE is convinced that its solar programme for commercial roof surfaces will drive numerous environmental initiatives in California, including the Million Solar Roofs Programme that subsidises PV projects up to 2017. The SCE initiative supports the programme of the federal state against climate change as well as the programme for renewable energies that stipulates that until 2010 about 10% of the energy demand is to be covered by regenerative sources.

California Leads the American Photovoltaics Market

The US photovoltaics market holds immense growth potential. California, leading in matters of solar power, has through its ambitious subsidising policies created the largest market in the USA in which in 2007 about 100 MW were installed. In comparison: in Germany just over one gigawatt was installed last year. With the Californian solar initiative and its solar roof programme, Governor Schwarzenegger wants to achieve about 3,000 MW by 2017 and will provide subsidisation to approximately 3.2 billion US$. In New Jersey, the government wants to move away from traditional subsidisation models and still wants to achieve a 22.5% share of renewable energies in its electricity mix by 2020. Other states such as New York, Arizona, Colorado, Connecticut and Massachusetts have their own subsidisation programmes.

Solar-Thermal Power Plants in the Desert as Gigawatt Electricity Suppliers of the Future

But not only photovoltaics are considered as technology of the future in the USA, but also gigantic solar-thermal power plants are to cover the immense electricity demand of Americans. The big electricity suppliers in California have recently announced projects with a total output of over 2,000 MW. The current federal energy bill also envisages tax benefits for this kind of electricity generation, but the continuation of these is still uncertain. Although the gigawatt projects have been planned already, many years often pass before construction actually commences.

Solar Heating and Cooling still Little Developed, but New Market is Forming

Classical solar thermics boomed in the USA in the early 1980s, but heating and cooling with solar thermics has long not established itself in spite of mostly ideal conditions. The solar thermal industry remained rather small and was mainly used for heating pools. But now solar thermics is said to have reached a turning point in North America, the organisers of Solar Gigawatt emphasise and report annual growth rates of 50%. How political support of solar thermics can be structured and what it can achieve, will also be discussed at the symposium.

(Rolf Hug is Editor in Chief and Executive Director at Heindl Server GmbH. The author can be reached at rolf.hug@heindl.de).

8

Solar Energy Grid Integration Systems – Energy Storage (SEGIS-ES)

Dan Ton, Georgianne H Peek, Charles Hanley and John Boyes

This paper describes the concept of augmenting the SEGIS (Solar Energy Grid Integration Systems) with energy storage in residential and commercial applications. It explains the scope, prevalence and applications of SEGIS program. The paper also highlights the areas where further, PV-specific research and development can be carried off and also offers recommendations as to proceed with the same.

1. Executive Summary

In late 2007, the US Department of Energy (DOE) initiated a series of studies to address issues related to potential high penetration of distributed photovoltaic (PV) generation systems on our nation's electric grid. This Renewable Systems Interconnection (RSI) initiative resulted in the publication of 14 reports and an Executive Summary that defined needs in areas related to utility planning tools and business models, new grid architectures and PV systems configurations, and models to assess market penetration and the effects of high-penetration PV systems.

Source: www1.eere.energy.gov © Sandia Corporation. Reprinted with permission.

As a result of this effort, the Solar Energy Grid Integration Systems (SEGIS) program was initiated in early 2008. SEGIS is an industry-led effort to develop new PV inverters, controllers, and energy management systems that will greatly enhance the utility of distributed PV systems.

This paper describes the concept for augmenting the SEGIS Program with energy storage in residential and small commercial (≤100 kW) applications. Integrating storage with SEGIS in these applications can facilitate increased penetration of distributed PV systems by providing increased value to both customers and utilities. Depending on the application, the systems can reduce customers' utility bills, provide outage protection, and protect equipment on the load side from the negative effects of voltage fluctuations on the grid. With sufficient penetration, PV-Storage systems are expected to reduce emissions related to generation and will be critical to maintaining overall power quality and grid reliability as grid-tied distributed PV generation becomes more common.

Although electric energy storage is a well-established market, its use in PV systems is generally for stand-alone systems. The goal of SEGIS Energy Storage (SEGIS-ES) Program is to develop electric energy storage components and systems specifically designed and optimized for grid-tied PV applications. The Program will accomplish this by conducting targeted research and development (R&D) on the applications most likely to benefit from a PV-Storage system (*i.e.*, peak shaving, load shifting, demand response, outage protection, and microgrids) and developing PV-Storage technologies specifically designed to meet those needs. Designing optimized systems based on existing storage technologies will require comprehensive knowledge of the applications and the available storage technologies as well as modeling tools that can accurately simulate the economic and operational effect of a PV-Storage system used in that application.

This paper describes the scope of the proposed SEGIS-ES Program; why it will be necessary to integrate energy storage with PV systems as PV-generated energy becomes more prevalent on the nation's utility grid; and the applications for which energy storage is most suited and for which it will provide the greatest economic and operational benefits to customers and utilities. Because selecting and optimizing the storage technology for the application will be critical to the success of any PV-Storage system, this paper also provides a detailed summary of the various storage technologies and compares their relative costs and development status (*e.g.*, mature, emerging, *etc.*). Finally, the paper highlights the areas where further, PV-specific R&D is needed and offers recommendations as how to proceed with the proposed work.

2. Vision

The US infrastructure for electricity generation and delivery is undergoing a revolution that will lead to increased efficiency, improved reliability and power quality for customers, 'smart' communications to match generation and loads, and the development of distributed generation from local and renewable resources. The high penetration of PV and other renewable energy technologies will be enabled by developing managed, efficient, reliable, and economical energy storage technologies that will eliminate the need for back-up utility baseload capacity to offset the intermittent and fluctuating nature of PV generation. These dispatchable storage technologies will bring added benefits to utilities, homeowners, and commercial customers through greater reliability, improved power quality, and overall reduced energy costs.

3. Program Objective

This Program will develop advanced energy storage components and systems that will enhance the performance and value of PV systems, thereby enabling high penetration of PV-generated electricity into the nation's utility grids. Through its RSI initiative, the DOE Solar Energy Technology Program is identifying needs and developing technologies to facilitate the high penetration of distributed electricity generation. The need for improved energy storage has been highlighted as a key factor to achieving the desired level of PV generation.

The electrical energy storage industry is well established and offers a variety of products for vehicle, Uninterruptable Power Supply (UPS), utility-scale, and other applications. The design and development of storage products specifically for PV applications, however, is nearly nonexistent. Traditional PV-Storage systems have been for off-grid applications that required some amount of autonomy at night and/or during cloudy weather. The objective of this Program is to develop energy storage systems that can be effectively integrated with new, grid-tied PV and other renewable systems and that will provide added value to utilities and customers through improved reliability, enhanced power quality, and economic delivery of electricity.

4. Program Scope

In late 2007, DOE began a series of studies to address issues related to the potential high penetration of distributed PV generation systems on our nation's electric grid. The RSI initiative resulted in the publication of 14 reports and an Executive Summary that defined needs in areas related to utility planning tools and business models, new grid architectures and PV systems configurations, and models to assess market penetration and the effects of high-penetration PV systems[1]. As a result of this effort, the SEGIS program was initiated in early 2008. SEGIS is an industry-led effort to develop new PV inverters, controllers, and energy management systems that will greatly enhance the utility of distributed PV systems.

SEGIS-ES is closely related to the SEGIS Program, a three-year program whose goal is to develop new commercial PV inverters, controllers, and energy management systems with new communications, control, and advanced autonomous features[2]. The heart of the SEGIS hardware, the inverter/controller, will manage generation and dispatch of solar energy to maximize value, reliability, and safety, as we move from 'one-way' energy flow in today's distribution infrastructure to 'two-way' energy and information flow in tomorrow's grid or microgrid infrastructure.

The applicable markets for the SEGIS Program[3] are defined in Table 1. The table shows the size of the PV system in watts, or power output. Storage systems are typically rated in terms of energy capacity (i.e., watt-hours) which is highly dependent on the application for which the storage is being used. These applications are discussed later in this document.

Table 1: Target Market Sectors for SEGIS PV Systems	
Residential	Less than 10 kW, single-phase
Small Commercial	From 10 to 50 kW, typically three-phase
Commercial	From 50 to 100 kW, three-phase

SEGIS-ES is focused on developing commercial storage systems for distribution-scale PV in the market sectors shown in Table 1; specifically, PV systems designed for applications up to 100 kW that can be aggregated into multi-megawatt systems. Integrating electric energy storage into homes or commercial buildings is also a key focus of SEGIS-ES. New storage systems developed under the Program will play an important role in the development of independent microgrids – either individual buildings or communities of buildings – so microgrid-scale storage, on the order of a megawatt of distributed generation, is within the scope of this effort. Storage systems developed through SEGIS-ES will interface with SEGIS products to further enhance PV system value and economy to customers. Products to be developed through SEGIS-ES include, but are not limited to, the following:

- Battery-based systems using existing technologies that are enhanced or specifically designed for PV applications including the development of PV-Storage hybrid systems;
- New energy storage system controllers that interface with SEGIS hardware to optimize battery use in order to obtain the best possible system efficiency and battery life;
- Non-battery storage systems (e.g., electrochemical capacitors [ECs], flywheels) designed specifically for PV applications; and
- New devices that integrate into building infrastructure.

SEGIS-ES does not address the following:

- Development of PV modules.
- Development of new battery technologies, although collaboration with the DOE Office of Basic Energy Sciences Energy Frontier Research Centers' Funding Opportunity is encouraged.
- Utility-scale storage systems or storage at the level of large distribution feeders. Although these efforts are key to achieving high penetration of distributed generation, they will be addressed through other Program activities.
- PV inverters or related power conditioning devices.
- Non-solar-related storage system development, smart appliances, or utility portals.

5. The Need for Energy Storage in High-Penetration PV Systems

PV systems are only a small part of today's electric infrastructure and have little effect on the overall quality or reliability of grid power. Nevertheless, state and federal efforts are currently underway to greatly increase the penetration of PV systems on local and regional utility grids to achieve goals related to emissions reduction, energy independence, and improved infrastructure reliability. When PV penetration reaches high enough levels (*e.g.*, 5 to 20% of total generation) however, the intermittent nature of PV generation can start to have noticeable negative effects on the entire grid. Figure 1 illustrates the transient nature of PV generation as clouds pass over a typical residential system during the course of a day. Both the magnitude and the rate of the change in output are important: in mere seconds, the PV system can go from full output to essentially zero and back again. At high levels of PV penetration, this intermittency can wreak havoc on utility operations and on load-side equipment due to fluctuations in grid voltage and power factor. Stated simply, fluctuations on this scale will not be allowable.

Figure 1: Measured and Modeled PV System Output on a Day with Frequent Passing Clouds

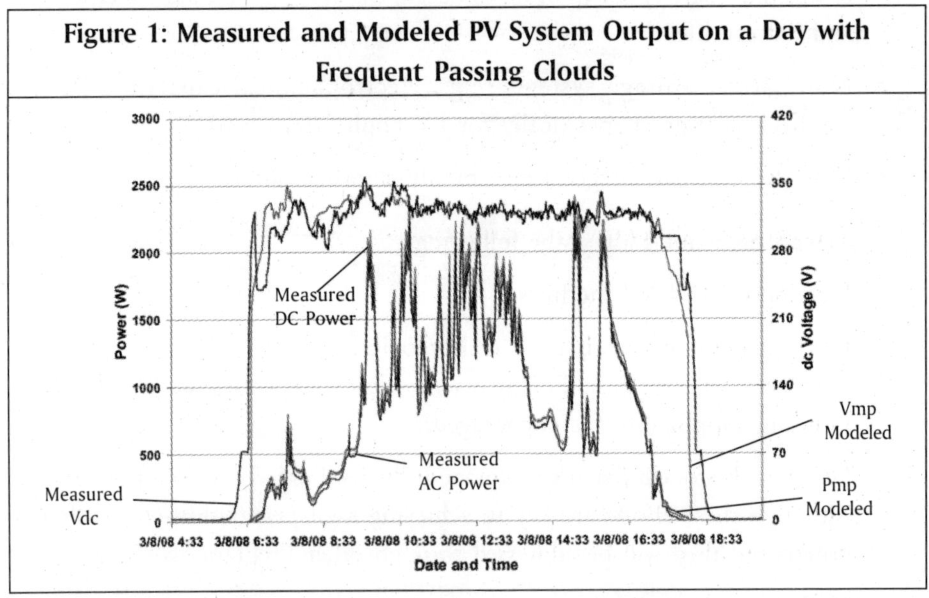

The distributed nature of PV can help to mitigate the negative consequences of high PV penetration to some degree; over large regions the effects of intermittent generation on the grid will be less noticeable. Nevertheless, utilities will still need to address worst-case possibilities. When transients are high, area regulation will be necessary to ensure that adequate voltage and power quality are maintained. When PV generation is low, some type of back-up generation will be needed to ensure customer demand is met. Additionally, because most utilities require an amount of 'spinning reserve' power that typically is equal to the power output of the largest generating unit in operation, the amount of spinning reserve necessary will increase with the amount of distributed PV generation that is brought online.

As the graph in Figure 2 illustrates, high PV penetration may reduce intermediate fossil fuel generation but, without storage, will do little or nothing to reduce a utility's overall conventional generation because of higher spinning reserve requirements.

Figure 2: The Need for Additional Spinning Reserve or Storage to Back up Intermittent PV Generation at Increasing Levels of Penetration

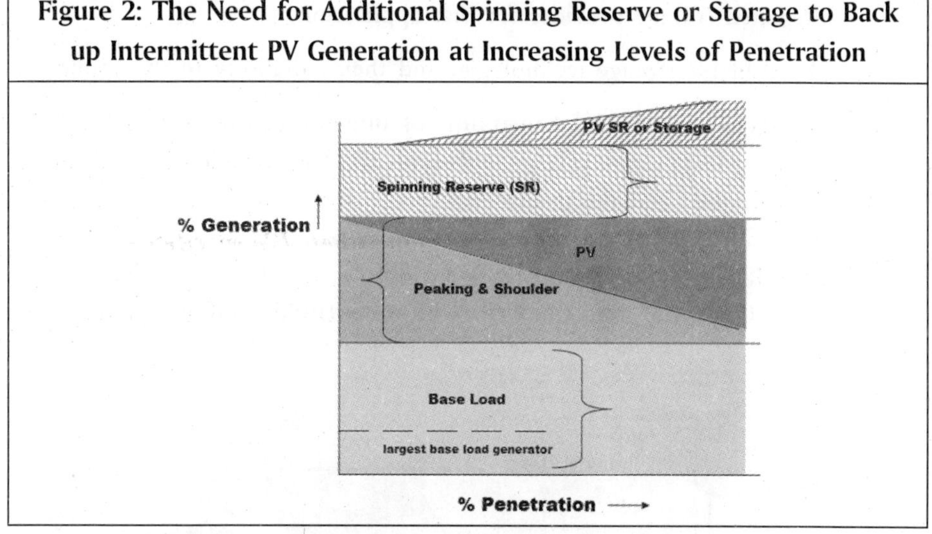

The utility grid as a whole must evolve in several ways to accommodate increased penetration of PV and other distributed and intermittent electricity generation sources including improved flexibility, better load management, integration of storage technologies, and even limited curtailment for extreme events. Several efforts are underway to define the next generation grid infrastructure that include the areas listed above[4]. A recent study that specifically focused on the current grid and high-penetration PV called energy storage the 'ultimate solution' to allow intermittent sources to address utility baseload needs and stated that "a storage system capable of storing substantially less than 1 day's worth of average demand could enable PV to provide on the order of 50% of a system's energy[5]."

This paper focuses on incorporating storage as part of this overall 'systems' solution. Successfully integrating energy storage with distributed PV generation in grid-connected applications involves much more than selecting an adequately sized system based on one of the many commercially available technologies. The complexity of a grid-integrated PV-Storage system is illustrated in Figure 3, which shows SEGIS-based generation integrated with electrical energy storage for a residential or small-commercial system. Optimal integration of storage with grid-tied PV systems requires a thorough understanding of the following:

- The application for which the storage is being used and the benefits integrated storage provides for that application;
- The available storage technologies and their suitability to the application;
- The requirements and constraints of integrating distributed generation and electrical energy storage with both the load (residential, commercial, or microgrid) and the utility grid;

Figure 3: The Relationship Between SEGIS, Electric Energy Storage, the Customer, and the Utility in an Optimal Configuration[6]

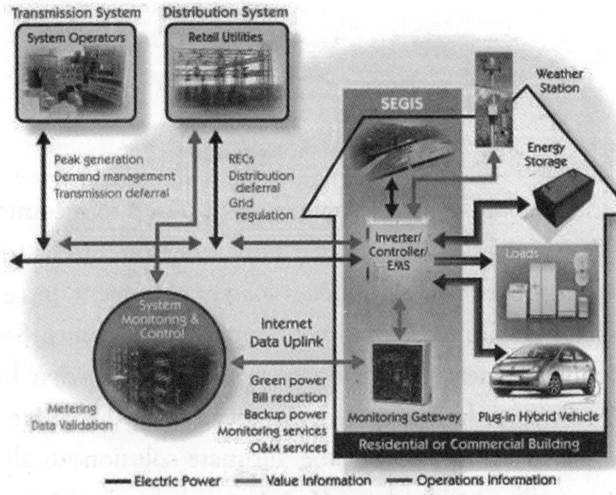

- The power electronics and control strategies necessary for ensuring that all parts of the grid-connected distributed generation and storage system work; and
- The requirements to provide service to the load and to maintain or improve grid reliability and power quality.

6. Applications of Energy Storage in High-Penetration PV Systems

Integrated PV-Storage systems provide a combination of financial, operational, and environmental benefits to the system's owner and the utility through peak shaving and reliability applications[7]. Peak shaving, load shifting, and demand response are variations on a theme—supplying energy generated at one time to a load at some later time. The rate structure and interactions between the utility and the customer determine which application is being addressed. Outage protection and power quality control increase the reliability of the electric grid and are not as subject to regulatory and rate-based actions. All of these sub-applications are summarized below.

Peak Shaving: The purpose of this application is to minimize demand charges for a commercial customer or to reduce peak loads experienced by the utility. Peak shaving using PV-Storage systems requires that the PV provide all required power above a specified threshold and, if PV is not available, adequate energy storage to fill the gap. Failure to peak shave on one day can have severe economic consequences in cases where customers' rates are based on monthly peak demand. Thus, reliability of the PV-storage system is a key element. To implement peak shaving, the system controller must be able to dispatch power from the storage system if PV is unavailable to meet the load.

Load Shifting: Technically, load shifting is similar to peak shaving, but its application is useful to customers purchasing utility power on a time-of-use (TOU) basis. Many peak loads occur late in the day, after the peak for PV generation has passed. Storage can be combined with PV to reduce the demand for utility power during late-day, higher-rate times by charging a storage system with PV-generated energy early in the day to support a load later in the day. Pacific Gas and Electric (PG&E) offers the experimental rate structure shown in Table 2 for residential customers. In this schedule, peak rates apply between 2 p.m. and 7 p.m. on weekdays and super-peak rates apply between 2 p.m. and 7 p.m. for no more than 15 days in a calendar year during critical events (as designated by the independent system operator, or ISO) and emergencies. Thus, customers with a PV-Storage system could use PV to charge the storage device earlier in the day (*i.e.*, during peak insolation) and then use the storage system to supply all or part of the load when peak or super-peak rates are in effect. With rate structures such as these in effect, a PV-Storage system could potentially provide significant economic benefits to residential and small commercial customers.

Table 2: PG&E Rate Schedule E-3—Experimental Residential Critical Peak Pricing Service[8]

Total Energy Rates ($/kWh)	Super Peak	Peak	Off Peak
Summer Baseline Usage	0.67439	0.23096	0.08039
Winter Baseline Usage	0.50997	0.31197	0.10497

Demand Response: Demand response is fast becoming a viable load management tool for electric utilities. Demand response allows the utility to control selected high-load devices, such as Heating, Ventilation, and Air Conditioning (HVAC) and water heating, in a rolling type of operation during high-demand periods. Utility rate structures are currently changing to accommodate this new operational strategy by reducing rates for customers who choose to be included in the demand response program. For both residential and small commercial customers, using an appropriately sized PV-Storage system should allow the implementation of demand response strategies with little or no effect on local operations. While both residential and commercial PV-Storage systems have the inherent capability to manage demand response requirements, control systems capable of reacting to demand response demands will have to be developed. Specifically, control systems must dispatch the PV-Storage system as necessary to manage the loads curtailed by the demand response program. Consequently, at least one-way communication with the utility may be required.

Outage Protection: An important benefit of a PV-Storage system is the ability to provide power to the residential or small commercial customer when utility power is unavailable (*i.e.*, during outages). To provide this type of protection it is necessary to intentionally island the residence or commercial establishment in order to comply with utility safety regulations designed to prevent the back-feeding of power onto transmission and distribution (T&D) lines during a blackout. Islanding requires highly reliable switching equipment for isolating the local loads from the utility prior to starting up local generation. Islanding capability whether utility- or customer-controlled, is mutually beneficial to both the utility and the customer because it allows the utility to shed loads during high demand periods while protecting the customer's loads if the utility fails. To realize the full benefit of these capabilities, however, new controllers are needed to respond to both the utility and customer needs. Additionally, new regulations will be needed to define how these controllers will be managed to benefit both the utility and the customer safely.

Grid Power Quality Control: In addition to outage protection, power quality ensures constant voltage, phase angle adjustment, and the removal of extraneous harmonic content from the electric grid. On the customer side, this function is currently supplied by UPS devices. A UPS must sense, within milliseconds, deviations in the AC power being supplied and then take action to correct those deviations. A common deviation is a voltage sag in which the UPS supplies the energy needed to return the voltage to the desired level. UPS functions can be added to PV-Storage systems in the power conditioning system by designing it to handle high power applications and including the necessary control functions. UPS functionality can be combined with peak shaving capability in the same system.

Microgrid: The incorporation of microgrids into the larger grid infrastructure is expected to become an increasingly important feature of future distribution systems. Microgrids have the potential to significantly increase energy surety[9]. Renewable generation and energy storage are essential to achieving highly sustainable, highly reliable microgrids. When operating separately from the local utility (*i.e.,* when 'islanded'), microgrids with PV-Storage systems will use PV-generated electricity to supply power to the load. Energy storage is essential to ensure stable operation of the microgrid by managing load and supply variations and for keeping voltage and frequency constant. Successful integration of microgrids that include PV-Storage systems to the larger utility grid infrastructure will provide many operational benefits to utilities and customers but will require a high level of system control, a detailed knowledge of the load(s) being served, and thoughtful design of the PV-Storage system.

Current and future applications that can be addressed by integrating energy storage with distributed PV generation are summarized in Table 3.

Table 3: Applications for Storage-Integrated PV

Residential			
Homeowner-Owned Systems		**Utility-Owned Systems**	
Current	**Future**	**Current**	**Future**
• Save for energy for evening use in TOU operations • Back-up power (UPS)	• With time of day residential rates, load shifting • Lower cost than utility • Smart grid interface	• Solar community – ride through during cloud cover • Distributed generation • Congestion reduction	• Smart grid applications (e.g., distributed energy management microgrid islanding peak shaving/shifting) • High penetration ramp control (short-term spinning reserve) • Emission reduction carbon credits (with high penetration)
Commercial			
Business-Owned Systems		**Utility-Owned Systems**	
Current	**Future**	**Current**	**Future**
• Peak shaving to reduce TOU and/or demand changes • Power quality and UPS	• Carbon Credits • Microgrid generation and islanding • Smart grid/building management interfaces	• Distributed generation • Congestion reduction • Improved power quality	• Microgrid generation and islanding • Emission reduction, carbon credits (with high penetration)

The economic benefits that can be realized from PV-Storage systems are a function of the application, the size of the system, the sophistication of the system's electronic control equipment, the customer's rate structure, and the utility's generation mix and operating costs. Systems that include UPS features are expected to mitigate the costs of power quality events and outages. Results of a recent study also suggest that adding PV generation to a planned UPS installation is attractive because of the synergy between PV and storage in the UPS market. In other words, sites where customers have already decided to purchase load protection via energy storage may be an attractive near-term target for PV developers[10].

In general, however, most financial benefits will result from reduced peak-demand and TOU charges for consumers and the avoided costs of maintaining sufficient peak and intermediate power generating capability plus spinning reserve for utilities. By facilitating an optimal mix of generation options, it is expected that the cost to the utility of adding additional generating capacity and the associated T&D equipment can be reduced as can the costs associated with upgrading existing T&D equipment to meet new demand. Additional future financial benefits could accrue to the end user by selling power back to the utility and to the utility by selling carbon credits realized by aggregating PV generation as a market commodity. Ideally, rate structures for PV-Storage systems could be designed to benefit both system owners and utilities. To fully realize all of the potential economic benefits will, however, require an advanced control system that includes communications between the utility, the PV-Storage system, and (possibly) the customer.

Finally, at high levels of penetration, PV systems offer significant environmental benefits. One such benefit is that they create no emissions while generating electricity. Another is that they can be installed on rooftops and on undesirable real estate, such as brown fields, which can reduce a utility's need to acquire land for building new large-scale generating facilities, the associated local opposition to such acquisitions, and the environmental consequences of large-scale industrial construction. Adding electrical energy storage to distributed PV generation also produces no emissions and, by allowing PV-generated electricity to be used at times when PV would not normally be available (e.g., at night or when it is cloudy), allows greater benefits to be realized than with PV systems alone.

7. Current Electrical Energy Storage Technologies and R&D

Energy storage devices cover a variety of operating conditions loosely classified as 'energy applications' and 'power applications'. Energy applications discharge the stored energy relatively slowly over a long duration (*i.e.*, tens of minutes to hours). Power applications discharge the stored energy quickly (*i.e.*, seconds to minutes) at high rates. Devices designed for energy applications are typically batteries of various chemistries. Power devices include certain types of batteries, flywheels, and ECs. A new type of hybrid device, the lead-carbon asymmetric capacitor, is currently being developed and is showing promise as a device that may be able to serve both energy applications and power applications in one package. Figure 4 shows several battery and capacitor technologies in relation to their respective power/energy capabilities[11]. The traditional lead-acid battery stands as the traditional benchmark. The plot shows that significantly greater energy and power densities can be achieved with several rechargeable battery technologies.

Figure 4: Specific Power vs. Specific Energy of Several Energy Storage Technologies[12]

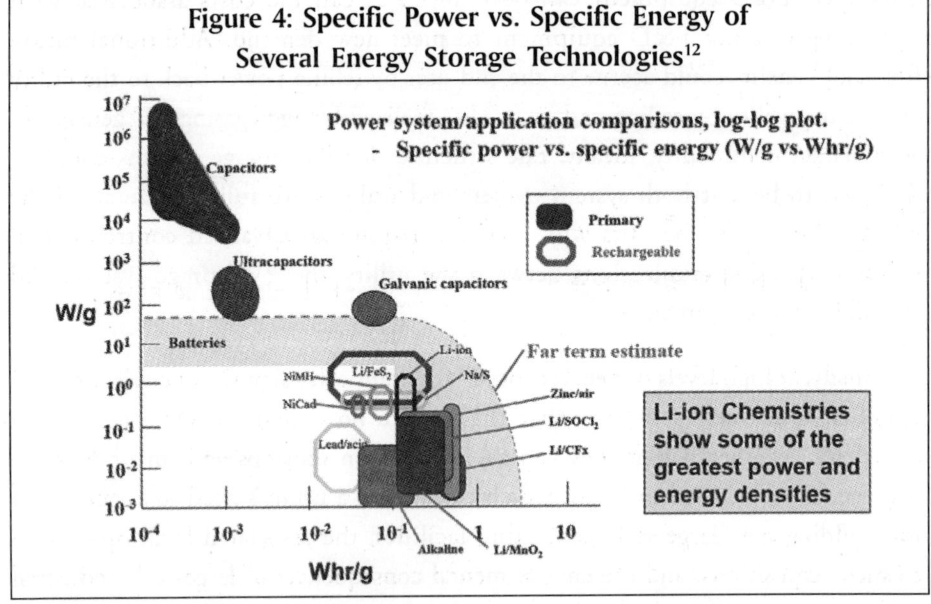

To date, the advantages of lead-acid technology, such as low cost and availability, have made it the default choice for energy storage in most PV applications. Indeed, new developments in Valve-Regulated Lead-Acid (VRLA) technology may revolutionize this well-established technology. A number of lead-acid battery manufacturers, such as East Penn in the US and Furukawa in Japan, are manufacturing prototype batteries for Hybrid Electric Vehicles (HEVs) that promise to overcome the main disadvantages of VRLA batteries by using special carbon formulations in the negative electrode. The added carbon inhibits hard sulfation, which minimizes or eliminates many common failure mechanisms (e.g., premature capacity loss and water loss). In cycling applications, the new VRLA technology could dramatically lower the traditional battery energy costs by increasing cycle life, efficiency, and reliability.

Traditionally, Nickel-Cadmium (NiCd) batteries have been the replacement for lead-acid, but due to various operational and environmental issues, industry is moving away from this technology as newer and better technologies are developed. Indeed, even in the portable electronics market lithium-ion (Li-ion) batteries are rapidly replacing NiCd. Additionally, a new Li-ion technology, the Li-iron phosphate (Li-FePO) cell, is rapidly becoming a prime contender for the next generation of HEV batteries, replacing existing nickel-metal hydride (Ni-MH) technology. This Li-ion technology is proving to be much safer than the previous generation and is capable of higher power levels, which makes it a better candidate for HEV applications. A lesser known technology, sodium/nickel-chloride (Na/NiCl), has been developed by Zebra Technologies in Europe for motive applications and is currently being considered for some stationary applications, such as peak shaving, in the US. Other advanced battery technologies (*e.g.*, sodium/sulfur, or Na/S) are currently targeting utility-scale (> 1MW) stationary applications. Although these technologies are not currently being considered for use in the smaller applications discussed here, future advances in the technologies themselves may increase their technical and economic viability for such applications. Table 4 summarizes the battery technologies that have been identified as potential candidates for integration with grid-tied distributed PV generation in residential and small commercial systems.

Table 5 provides a summary of non-battery technologies that may be integrated with grid-tied distributed PV generation. Although they are still in the early commercial stage of development, hybrid lead-carbon asymmetric capacitors are also targeting the peak-shaving market and low-speed flywheels are currently being used in many UPS applications. ECs are ideal for high-power, short-duration

Table 4: Battery Technologies for Electric Energy Storage in Residential and Small-Commercial Applications

Technology	Advantages	Disadvantages	Commercial Status	Current R&D	Applications
Flooded Lead-acid	• Cost effective • Mature technology • Relatively efficient	• Low energy density • Cycle life depends on battery design and operational strategies when deeply discharged • High maintenance • Environmentally hazardous materials	• Globally commercial • Over $40B in all applicants • Estimated $1B in utility applications worldwide	Focused on reducing maintenance requirements and extending operating life	• Motive power (forklifts, cars, etc.) and deep-cycling staionary applications • Back-up power • Short-duration power quality • Short-duration peak reduction
VRLA	• Cost effective • Mature technology	• Traditionally have not cycled well • Have not met rated life expectancies	• Globally commercial • Over $40B in all applications • Estimated $1B in utility applications worldwide	Improving cycle-life and extending operating life, such as using carbon-enhanced negative electrodes	• Limited motive power applications (e.g., electric wheelchairs) • Back-up power • Short-duration power quality • Short-duration peak reduction
NiCd	• Good energy density • Excellent power delivery • Long shelf life • Abuse tolerant • Low maintenance	• Moderately expensive • "Memory Effect" • Environmentally hazardous materials	• Globally commercial • Over $1B in all applications • Over $50M in utility applications worldwide	None identified	• Aircraft cranking, aerospace, military and commercial aircraft applications • Utility grid support • Stationary rail • Telecommunications back-up power • Low-end consumer goods
NiMH	• Good energy density • Low environmental impact • Good cycle life	• Expensive	• Globally commercial for small electronics • Emerging market for larger applications	Bipolar design	• EVs, HEVs • Small, low-current consumer goods
Li-ion	• High energy density • High efficiency	• High production cost • Scale-up proving difficult due to safety concerns	• 50% of global small portable market	Batteries for use in EVs and HEVs are currently being developed	• Small consumer goods
Li-FePO[4]	• Safer than traditional Li-ion • High power density • Lower cost than traditional Li-ion	• Lower energy density than other Li-ion technologies	• High-volume production began in 2008	Focused on improving performance and safety systems	• Small consumer goods and tools • EVs, HEVs

Contd...

Contd...

Na/S	• High energy density • No emissions • Long calendar life • Long cycle life when deeply discharged • Low maintenance • Integrated thermal and environmental management	• Relatively high cost • Requires powered thermal management (heaters) • Environmentally hazardous materials • Rated output available only in 500-kW/600-kWh increments	• Recently commercial (2002) in Japan • Estimated $0.4B in utility/industrial applications worldwide	Focused in increasing manufacturing yield and reducing cost	• Utility grid-integrated renewable generation support • Utility T&D system optimization • Commercial/industrial peak shaving • Commercial/industrial backup power
Zebra Na/Nicl	• High energy density • Good cycle life • Tolerant of short circuits • Low-cost materials	• Only one manufacturer • High internal resistance • Molten sodium electrode • High-operating temperature	• Globally commercial for traction applications	Focused on cost reduction and systems for stationary applications	• EVs, HEVs and locomotives • Peak shaving
Vanadium Redox	• Good cycle life • Good AC/AC Efficiency • Low temperature/low pressure operation • Power and energy are independently scaleable	• Low energy density	• Commercial production since 2007	Focused on cost reduction	• Firming capacity of renewable resources • Remote area power systems • Load management • Peak shifting
Zinc/ Bromine	• Low temperature/low pressure operation • Low maintenance • Power and energy are independently scaleable	• Low energy density • Requires stripping cycle • Medium power density	• Emerging commercial products	Focused on system integration	• Back-up power • Peak shaving • Firming capacity of renewables • Remote area power • Load management

Table 5: Non-Battery Technologies for Electric Energy Storage in Residential and Small-Commercial Applications

Storage Type	Advantages	Disadvantages	Commercial Status	Current R&D	Applications
Lead-carbon asymmetric capacitors (hybrid)	• Rapid recharge • Deep discharge • High power delivery rates • Long cycle life • Low maintenance	• Low energy density than batteries • Lower power density than other ECs	• Non-commercial prototypes	Laboratory prototypes Field demonstration planned FY08 in NY	• Peak shaving • Grid buffering
Electro-chemical Capacitors	• Extremely long cycle life • High power density	• Low energy density • Expensive	• Commercialized in US, Japan, Russia, and EU, emerging elsewhere • Over $30 million in all applications • $5 million in utility applications by 2006	Devices with energy densities over 20kWh/m³ are under development	• HEVs • Portable electronics • Utility power quality • T&D stability
Flywheels	• Low maintenance • Long life • Environmentally inert	• Low energy density • High cost	• Commercialized in US, Japan, Europe, emerging elsewhere • Projected to sell over 1,000 systems per year, estimated rated capacity of 250 MW • Retail value exceeding $50 million by 2006	Focused on low cost commercial flywheel designs for long duration operation	• Aerospace • Utility power quality • T&D stability • Renewable support • UPS • Telecommunications

applications because they are capable of deep discharge and have a virtually unlimited cycle life. Because of these advantages a great deal of research is being focused on developing ECs that can be used for small-scale stationary energy storage. Any of these battery or non-battery technologies may be appropriate for residential and small-commercial integrated PV and storage systems in the near future.

In addition to the efforts of the technology manufacturers, DOE, through several program offices, is conducting research and executing pilot programs to improve the utilization of electrical energy storage for stationary applications. Since the late 1970s, the DOE Energy Storage Systems Program has worked with the utility industry to develop stationary energy storage systems for utility applications. In the 1990s, the Program shifted its focus from developing advanced storage technologies to include an emphasis on integrating storage devices with power electronics and communications equipment for use in specific applications. Over the last decade, the Program has gained valuable practical experience by partnering with storage technology manufacturers, power electronics and monitoring equipment manufacturers, systems integrators, electric utilities, and their customers to demonstrate integrated electric energy storage systems of all types and sizes. Lessons learned from the Program's demonstrations and research provides uniquely applicable experience for successfully incorporating electric energy storage with distributed PV generation.

The DOE Vehicle Technologies Program, in partnership with the automotive industry, manages and conducts research on battery technologies for EVs and HEVs (e.g., lithium-aluminum-iron-sulfide, Ni-MH, Li-ion, and lithium-polymer). Li-ion systems come closest to meeting all of the technical requirements for vehicle applications, but they face four barriers: calendar life, low-temperature performance, abuse tolerance, and cost. Technology advances that address these barriers will have direct applicability to PV-Storage systems for stationary applications.

Finally, the Office of Basic Energy Sciences conducted a comprehensive workshop on April 2-4, 2007 that set R&D priorities for improving the energy density of several storage technologies. Principal barriers identified at the workshop were related to reducing cost, increasing power and energy density, lengthening lifetime, increasing discharge times, improving safety, and providing reliable operation through one to ten thousand rapid charge/discharge cycles. Several associated R&D efforts are underway. One such effort is the announcement of a funding opportunity to establish Energy Frontier Research Centers (EFRCs) specifically focused on "addressing fundamental knowledge gaps in energy storage[13]." The SEGIS-ES program will be closely coordinated with any developments to come from these programs.

8. The Costs of Electrical Energy Storage

Both current and projected costs for battery and other storage systems are related to first-costs and are based on the overall energy capacity of those systems. Table 6 shows the current and projected first (or capital) costs of energy storage systems based on technologies identified as suitable for residential and small-commercial PV-Storage systems. This table was compiled from the results of a literature review and discussions with technology leaders at national laboratories and in industry.

Table 6: Energy Storage System Capacity Capital Costs[14,15,16,17,18,19,20,21]

Technology	Current Cost ($/kWh)	10-yr Projected Cost ($/kWh)
Flooded Lead-acid Batteries	$150	$150
VRLA Batteries	$200	$200
NiCd Batteries	$600	$600
Ni-MH Batteries	$800	$350
Li-ion Batteries	$1,333	$780
Na/S Batteries	$450	$350
Zebra Na/NiCl Batteries	$800*	$150
Vanadium Redox Batteries	20 kWh=$1,800/kWh; 100 kWh =$600/kWh	25 kWh=$1,200/kWh 100 kWh=$500/kWh
Zn/Br Batteries	$500	$250/kWh plus $300/kW†
Lead-carbon Asymmetric Capacitors (hybrid)	$500	<$250
Low-speed Flywheels (steel)	$380	$300
High-speed Flywheels (composite)	$1,000	$800
Electrochemical Capacitors‡	$356/kW	$250/kW

* €600/kWh.

† The battery system includes an integrated PCS; the PCS price will vary with the rated system output.

‡ Electrochemical capacitors are power devices used only for short-duration applications. Consequently, their associated costs are shown in $/kW rather than $/kWh.

Recent increases in the prices of materials for existing battery technologies, such as lead, have led to increased system costs. These trends are likely to continue, possibly driving the prices for established technologies even higher. Unless noted, the system costs include the storage device and the power conditioning system necessary for turning DC output from the storage device into 60-Hz AC power suitable for delivery to the load. For these systems, first costs will be lowered by combining the power electronics for both the PV and storage components.

Determining life-cycle costs depends on a number of factors related to system design, component integration, and overall use. Accurate prediction of life-cycle costs also depends on developing reasonably predictive models for PV-integrated storage. More modeling and analytical work are needed to determine the incremental Levelized Cost of Energy (LCOE) that storage will bring to PV systems and the incremental value of increased benefits.

9. Summary of Key R&D Needs for PV-Storage Systems

Achieving high-penetration of PV-Storage systems on the nation's utility grid will require overcoming certain technological and economic obstacles. In addition to the specific gaps described below, the successful implementation of optimal small-scale PV-Storage systems will require further development, testing, and demonstration of complete systems of varying complexity and cost.

9.1. Storage Technologies

To meet the needs of SEGIS-based systems, it will be necessary to develop battery and other storage systems that, while based on current state-of-the-art, are enhanced or specifically designed for use with grid-tied PV systems. It should be noted that any advances in storage technology will be of value to grid-tied PV-Storage systems because they further the understanding of the technology, which facilitates selection of the most appropriate technology for the application and, ultimately, reduces the cost of the storage components.

As previously stated, the main R&D needs for storage technologies address the following aspects of their use:

- Increasing power and energy densities;
- Extending lifetimes and cycle-life;

- Decreasing charge-discharge cycle times;
- Ensuring safe operation; and
- Reducing costs.

Batteries also do not typically work effectively under partial state of charge (PSOC) conditions. PSOC operation occurs when a battery is less than fully discharged and then less than fully recharged before being discharged again. Current research into carbon-enhanced lead-acid batteries shows high potential for significantly improving PSOC operation. Nevertheless, PSOC operation is not fully understood for all battery chemistries. Charge and discharge profiles for grid-connected PV-Storage applications need to be tested on the most promising technologies. Further development and optimization of batteries of various chemistries to improve PSOC operation is also needed.

9.2. Control Electronics

To achieve long lifetimes, maximum output, and optimal efficiency from batteries, they must be charged and discharged according to the manufacturer's recommendations. For example, traditional lead-acid batteries require a long (multiple-hour), low-current finish charge to remove sulfation from the lead plates. If finish charging is not done properly, battery lifetime is shortened and capacity is reduced. This finish charge is very difficult to accomplish with only a PV-based generation source. Advanced battery management systems can be developed to address some of the charge/discharge issues. The US Coast Guard is sponsoring an effort to develop the Symons Advanced Battery Management System (ABMAS) for off-grid, PV-Storage-Generator hybrid systems[22]. Initial results using the ABMAS system show a 25% reduction in fuel use and improved battery charging and discharging profiles promising increased battery lifetime. Similar management systems are needed for grid-connected PV-Storage systems and applications.

Energy storage devices themselves (batteries, flywheels, etc.) do not discharge power with a 60-Hz AC waveform (nor can they be charged with 60-Hz AC power); a power conditioning system is necessary to convert the output. The DOE Solar Energy Program is currently developing integrated power conditioning sytems for PV systems under the SEGIS initiative. These systems include inverters, energy management systems, control systems, and provisions for including energy storage. It is anticipated that charging and discharging control algorithms for different battery technologies will be included in the SEGIS control package. In the case of lead-acid and NiCd batteries, this will be relatively straightforward. Other technologies (e.g., Li-ion, vanadium redox, and Zn/Br batteries or flywheels) require more complex safety and control systems. These systems are typically sold by the battery manufacturer as part of an integrated, 'plug-and-play' energy storage system that includes the storage device, an inverter, and proprietary control and safety systems. To achieve the most economical total system using these technologies, SEGIS system manufacturers and manufacturers of these energy storage products could cooperate to design a fully integrated product with minimal duplicated functionality.

9.3. Comprehensive Systems Analysis

Successful development of SEGIS-based PV-Storage systems will require comprehensive systems analysis, including economic and operational benefits and system reliability modeling. Systems must be analyzed based on the requirements of the application and the analysis should include an investigation of all of the possible storage technologies suitable for use in the application and the operational/cost/benefits tradeoffs of each. This analysis must include a methodology for determining the life-cycle costs of PV-Storage systems using conventional industry metrics. This methodology will be used to determine benefit/cost tradeoffs for specific applications and system configurations.

Software-based modeling and simulation tools represent a key component of successful systems analysis. PV system designers use various models to evaluate the needs for and effects of various technologies. The system-level modeling software packages that are currently available to designers include: Solar Advisor Model (SAM), Hybrid Optimization Model for Electric Renewables (HOMER), PV Design Pro, and HybSim. For the most part, these models do not accommodate storage well. HybSim, funded by DOE's Energy Storage Systems Program, focuses on integrating storage, diesel generation, and wind or PV-generation. Ideally, models and simulation tools for grid-tied PV-Storage systems will be able to accomplish the following:

- Fully evaluate the benefits of a given PV-Storage system by modeling solar energy production, building loads, and energy storage capabilities relative to capital cost, maintenance, and the real-time cost of alternate energy sources (utility power).
- Accurately simulate residential, commercial, and utility systems and provide recommendations for how to operate, dispatch, and control the PV-Storage system to optimize its economic performance under various loads and rate structures.
- Provide detailed models of the interrelationships between the various system components and operating parameters including the physical relationships, operating rules, regulations, and business decision-making criteria to aid in comprehensive systems analysis and to identify relationships that might create unexpected vulnerabilities or provide additionally robustness.

10. Summary – The Path Forward

To address the technology gaps described above and to ensure that grid-tied PV-Storage systems meet the needs of customers, utilities, and all other stakeholders, a three-pronged approach is recommended:

- Comprehensive systems analysis and modeling;
- An industry-led R&D effort focused on commercialization of new integrated systems; and
- Development of appropriate codes and standards that facilitate broader market penetration of PV-Storage systems and address all related safety concerns.

These three aspects of SEGIS-ES are discussed in greater detail below.

10.1. Systems Analysis and Modeling

The RSI studies resulted in a series of reports that addressed the myriad issues related to high penetration of PV on utility infrastructure and business models, technical system design, and economic effects. A similar set of studies is proposed to fully investigate the role of energy storage in this environment. These analytical studies will include developing new modeling tools and will address the following:

- Developing models that explore several aspects of PV-Storage system integration, including system technical performance optimization; grid operational performance, stability, and reliability; cost/benefits; life-cycle costs; *etc.* Models will also address advantages and disadvantages of distributed versus aggregated storage systems (*e.g.*, community-scale *vs.* residential), and the integration of PV-Storage systems with building loads, operating rules, and regulations and business decision-making criteria to identify relationships that might create unexpected vulnerabilities or provide additional robustness.
- Investigating the integration of Energy Management Systems (EMS) with PV-Storage systems to optimally manage power for commercial facilities, including developing predictive algorithms for loads and PV output to effectively manage storage.
- Exploring the role and potential for Plug-in Hybrid Electric Vehicles (PHEVs) to provide grid and PV generation support. Using PHEVs for grid support or as energy storage devices to support residential/small-commercial distributed PV generation presents unique challenges for system integrators because they are mobile devices. Consequently, we recommend investigating how PHEV-based storage can best be aggregated to support distributed PV generation and determining the operational requirements and system specifications necessary for doing so.

10.2. Partnered Industry Research and Development

An industry-led effort will be initiated to strengthen ties between manufacturers and installers in the storage industry with appropriate partners and stakeholders in the PV industry (including utilities), to achieve the following goals:

- Development of new components and integrated PV-Storage systems for grid-connected applications by identifying the requirements and constraints of integrating distributed generation and electrical energy storage with both the load (residential, commercial, or microgrid) and the utility grid. This effort will include developing the power electronics and control strategies necessary to ensure that all parts of the grid-connected distributed generation and storage system work as expected to provide service to the load and to maintain or improve grid reliability and power quality.

- Test and verification of promising battery technologies using charge/discharge profiles specifically designed for grid connected PV-Storage applications in order to develop and optimize the PSOC operation of these battery chemistries.
- Providing the training (and in some cases cross training) necessary for successfully installing, operating, maintaining, and troubleshooting these highly integrated systems.

10.3. Codes and Standards Development

Ultimately, high levels of penetration of grid-tied distributed generation and storage will affect the utility grid and those who use it in many significant ways. Consequently, codes standards, and regulations for integrating these systems with the grid will also be needed to facilitate this integration. Additionally, safety guidelines and regulations that specifically address the complexities of these systems will need to be developed and implemented. The development of this regulatory environment will be a concerted effort that will build on the current codes and standards infrastructures that exist for the PV, energy storage, construction, and utility industries, and will lead to a comprehensive set of guidelines that will facilitate the greater market penetration of PV-Storage systems.

Acknowledgement

This manuscript has been authored by Sandia Corporation under Contract No. DE-AC04-94AL85000 with the US Department of Energy. The United States Government retains and the publisher, by accepting the article for publication, acknowledges that the United States Government retains a non-exclusive, paid-up, irrevocable, world-wide license to publish or reproduce the published form of this manuscript, or allow others to do so, for United States Government purposes.

(Dan Ton is the Team Leader of the Technology Evaluation activity at the US Department of Energy's (DOE) Solar Energy Technologies Program. The author can be reached at dan.ton@ee.doe.gov,

Georgianne H Peek is researcher at Sandia National Laboratories. The author can be reached at ghpeek@sandia.gov,

Charles Hanley, Sandia National Laboratories. The author can be reached at cjhanle@sandia.gov, and

John Boyes is Manager at Energy Storage & Distributed Energy Resources, Sandia National Laboratories. The author can be reached at jdboyes@sandia.gov).

11. References

1. Kroposki, B.; R. Margolis, G. Kuswa; J. Torres; W. Bower; Ton, D., *Renewable Systems Interconnection: Executive Summary,* 2008. Downloadable at *http://www1.eere.energy.gov/solar/solar_america/pdfs/42292.pdf*
2. Ton, D.; Cameron, C.; Bower, W.; *Solar Energy Grid Integration Systems "SEGIS".* Concept Paper.
3. Ibid. [1]
4. Maire, J.; Von Dollen, D. *Profiling and Mapping of Intelligent Grid R&D Programs.* Report 1014600 to the IEEE Working Group o Distribution Automation. December 2006.
5. Denholm, P.; Margolis, R. Evaluating the Limits of Solar Photovoltaics (PV) in Electric Power Systems Utilizing Energy Storage and Other Enabling Technologies. April 2007.
6. Ibid. [1].
7. Manz, D.; Schelenz, O.; Chandra, R.; Bose, S.; de Rooij, M.; Bebic, J. *Enhanced Reliability of Photovoltaic Systems with Energy Storage and Controls.* RSI Study NREL/SR-581-42299. February 2008.
8. Pacific Gas and Electric Company (*www.pge.com/tariffs/ERS.shtml*)
9. Boyes, J.; Menicucci, D. "Energy Storage: The Emerging Nucleus". *Distributed Energy.* January/February 2007.
10. Hoff, T.; Perez, R.; Margolis, R. *Maximizing the Value of Customer-sited PV Systems Using Storage and Controls.*
11. Roth, E.P.; Doughty, D. *Thermal Response and Flammability of Li-ion Cells for HEV Applications.* SAND2005-1791P.
12. Ibid. [10].
13. *Basic Research Needs for Electrical Energy Storage.* Report of the Basic Energy Sciences Workshop for Electrical Energy Storage. July 2007.
14. Schoenung, S.; Eyer, J. *Benefit/Cost Framework for Evaluating Modular Energy Storage.* SAND2008-0978.
15. Tiax, LLC. *Energy Storage: Role in Building PV-based Systems.* Final report to DOE EERE. March 2007.

16. E-mail communications with Tom Hund of Sandia National Laboratories and Jim McDowall of SAFT America, Inc. March 14, 2008.

17. E-mail communication with Brian Beck of VRB Power Systems, Inc. April 11, 2008.

18. E-mail communications with Peter Gibson and Doug Alterton of Premium Power Corporation. April 15, 2008.

19. Dickinson, E.; Clark, N. "Development of High-performance Electrodes Containing Carbon for Advanced Batteries and Asymmetric Capacitors". DOE Energy Storage Systems Program FY08 Quarter 1 Report (October through December 2007). April 2008.

20. Eckroad, S.; Gyuk, I. EPRI-DOE Handbook of Energy Storage for Transmission & Distribution Applications. December 2003.

21. E-mail communication with Ib Olsen of Gaia Power Technologies, Inc. April 15, 2008.

22. Corey, G. "Optimizing Off-grid Hybrid Generation Systems". *EESAT 2005 Conference Proceedings.*

9

Solarscapes – A New Face for PV

Topher Donahue

The article highlights the idea of integrating photovoltaic into architectural structures by way of BIPV (Building Integrated Photovoltaic) in both new and existing structures. BIPV also minimizes the aesthetic concerns of adding PV to one's home.

If you count yourself among the people who love the idea of making their own clean energy, but balk at the idea of planting a pole mount in the middle of your backyard or covering your historic home's rooftop with high-tech PV modules, this fresh design strategy offers a solution.

The concept—aptly dubbed "solarscaping"—is a new take on building integrated photovoltaic (BIPV) systems. Unlike traditional BIPV systems that are designed into a structure from the blueprint phase, solarscapes work with both new and existing structures and can minimize the aesthetic concerns of adding PV to your home.

"Solarscaping is another avenue to facilitate our mission of helping people choose solar power," says Scott Franklin, president of Lighthousesolar, a solar system design

Source: www.homepower.com © Topher Donahue and Home Power Magazine. Reprinted with permission.

and installation company based in Boulder, Colorado. "The more options we can provide our customers, the greater chance we can meet their needs. These designs allow customers to get more function from their PV investment."

Though the idea of integrating PV into architectural structures is not new, Lighthousesolar is one of the first to offer a package option that also can be custom-fit to an application. Installation of a solarscape involves minimal time at the site. Structures are customized in the company's workshop in Boulder, delivered to the installation site in several large pieces, and assembled in one or two days. So far, the company has integrated PV systems into awnings, pergolas, carports, hot tub shades, and gazebos, and the potential is limitless, Franklin says. Fences, fountains, greenhouses, archways, and sunrooms are all good candidates, he adds.

What's in a Solarscape?

A solarscape features two elements. The first is a wooden or steel custom-made frame that is painted, trimmed, or finished to complement the home and landscaping.

The second component—the HIT (heterojunction with intrinsic thin layer) bifacial modules made by Sanyo—is key to the design's sleek look and increased energy production. These double-sided modules—which harvest solar energy from both the front and back faces—maximize power within a fixed amount of space. In an ideal setting, bifacial panels can be the most cost-effective modules based on dollars per watt.

Instead of the standard opaque module backsheet, Sanyo Double modules have glass-on-glass construction. Clear glass layers on both sides of the photovoltaic cells allow additional reflected solar energy to be captured from the back side of the module. The glass-on-glass construction adds aesthetic value as well by allowing some light to filter through the array. The subtle octagonal design of the silicon cells projects a soft light-and-shadow pattern on the surfaces beneath the array—similar to light filtering through the leaves of a large-leafed tree. Enough light passes through the modules to allow plants to grow underneath the array canopy while providing adequate protection for those seeking respite from the sun.

Behind Bifacial

Introduced in 1976, bifacial PV technology was most widely known for its use on spacecraft, including the International Space Station. Designed in response to the challenges of keeping monofacial modules oriented to the sun, bifacial arrays allow spacecraft more orientation options when orbiting the Earth. As an added bonus, the rear side of bifacial arrays can convert reflective radiation from the atmosphere into extra energy.

Over the past decade, bifacial PV technology has found its place on Earth, incorporated mainly in commercial applications such as awnings, street signs, bus stop shelters, and sound barriers. Recent innovations in optic technology have reduced the amount of silicon needed in a PV module and brought the cost of bifacial technology down to an earthly level, with a cost per watt about 20% higher than comparable single-sided modules. A lower price tag and opportunity for greater energy production has made bifacial PV a practical component for BIPV systems, as well as for use in systems that can earn performance-based incentives and renewable energy credits.

Unlike monofacial (single-sided) PV modules, the back side of a bifacial module generates power from light that is reflected off surrounding surfaces. Radiation from reflective surfaces—light-colored wood, metal roofing, concrete, white gravel, snow, or water—can increase the energy yield beyond the manufacturer's Standard Test Conditions (STC) output ratings. Sanyo estimates that the bifacial construction adds approximately 10% to module output when compared to a single-sided Sanyo HIT module in an angled installation, or as much as 34% in a vertically oriented installation.

"We know the bifacial panels have a strong power advantage, but it's hard to quantify because each installation has different characteristics," Franklin says. "Even Sanyo is hesitant to predict output because the numbers can vary significantly based on installation specifics."

A solar-electric carport.

Based on reports he has received from the field, Benjamin Collinwood, Sanyo's solar market development manager, estimates that typical production increases will be between 15% and 20%. He says that system results vary, depending upon individual site characteristics such as system design, location, and site albedo.

Solarscaping Solutions

Monofacial modules integrated into an awning serve a dual purpose: making electricity and providing shade for the windows.

Beyond the reflective radiation advantage, solarscapes can offer a much-needed solution for properties without a shade-free location for a roof- or ground-mounted array. Likewise, many buildings have less-than-optimal orientations for solar energy collection. Solarscaping can provide new and attractive options for array siting in many locations.

Homeowner Neil Cannon wanted to integrate PV into his newly built house in Eldorado Springs, Colorado, but he did not want to compromise the home's historic-looking design. Because tall trees shaded areas closer to the house, he needed a freestanding unit that could be located more than 100 feet away. Adding to the challenge, the structure had to be tall enough to avoid the shade of nearby trees.

Functional Design

Tucked in the Rocky Mountains at 8,236 feet above sea level, Scott Franklin's home occupies an idyllic spot in small-town Nederland, Colorado. Like some homes, its site is far from ideal for accommodating a roof-mounted solar installation. The house's solar access was limited by a northwest-southeast roofline and blocked by neighbors' trees. By building a detached office with an east-west orientation two years ago, Franklin created room for a 1.4 KW system.

Wanting to offset more of his family's household electricity consumption, Franklin explored other options. Knowing that a roof-mounted system on the home would be ineffective and a ground-mounted system would eat into the children's play area, he looked for a novel way to integrate an additional PV system.

"We'd lived in our home for six years and barely used our back, south-facing deck. We couldn't sit out there on a sunny day because it was either too bright or too hot," Franklin says. "An awning seemed like a natural solution for both problems".

So last spring, Franklin and the Lighthouse solar crew installed the company's first-built Power Awning—a custom-welded, 10- by 21-foot steel frame that incorporates fourteen 190 W bifacial modules—over the existing deck. The crew crafted the steel frame in the workshop and assembled the awning at the home over three days.

Contd...

Contd...

The picnic table beneath the awning has become the Franklins' favorite spot for both morning coffee and evening barbecues. During the mountain winters, some sunlight can pass through the array, preventing the dark, cold shadow created by a conventional awning and bringing some light into the home's interior.

Although the dark deck planks limit how much light is reflected to the underside of the array, the energy production of the awning's bifacial panels has exceeded Franklin's original expectations. On clear summer days, the net production from the Power Awning and the small roof-mounted system on the office is enough to spin the utility meter backward.

"The nicest thing is that the awning created a usable space for the family. Now we can sit out there and enjoy the view of the mountains," Franklin says. "The added bonus is that our electric bill made it down to zero".

The Power Awning cost about the same as a ground-mounted system but saved the crew from the time-consuming and backbreaking work of drilling through solid granite and running cables hundreds of feet to the inverter. A similarly sized roof-mounted system would have cost about 30% less than the Power Awning. The increased energy production of the bifacial modules helps to offset a portion of the higher cost. After federal and state rebates, the 2.66 KW awning system cost about $14,000.

The solution? A customized double carport roofed with 4.5 kilowatts (KW) of PV modules to meet 100% of the household's electricity needs, and several evacuated tube collectors for solar water heating. To keep with the look and feel of the home, the sides of the structure will be finished with rough-sawn lumber that resembles old barn siding.

A solar-electric awning over the Franklins' backyard deck provides additional energy generation and a comfortable space for outdoor activities.

"We're really charged about the idea," Cannon says. "It was really important that the structure be congruous with the landscape and the home, and this is a great compromise. Up close you'll be able to see the high-tech gear, but from far off in the distance, it'll blend in nicely".

Solarscapes offer several access and maintenance advantages over typical roof-mounted PV systems. The absence of roof-mounted arrays, for example, means roof maintenance and remodeling can be done without dismantling the solar-electric infrastructure. Increased airflow around the array will keep module operating temperatures lower and result in increased energy harvest.

Putting a Shine on PV

A prototype of Lighthousesolar's signature Power Awning at Franklin's home has won over several homeowners. All it took was one look at the awning's sleek design—a black steel frame inset with double-sided PV modules—to sell Marcus Luscher on the concept.

"When you first look at the awning, you don't even notice that it's a PV system. Only when you actually sit beneath it and see the PV panels do you realize that the awning is also generating power," Luscher says. "It's an attractive piece of architecture that is multifunctional." He plans to add a power awning above the deck at his home in Nederland, Colorado. The solarscape system will supplement his current nine-module PV system and help offset the electricity consumption of his newly purchased hot tub.

Double-sided Solutions

While bifacial PV awnings and carports are common in commercial BIPV installations, the concept is relatively new to residential projects. But solarscaping is gaining traction in home-scale installations because of potential aesthetic and versatility advantages when compared to traditional roof- or ground-mounted arrays.

Currently, Lighthousesolar delivers and installs customized units in Colorado and Texas, but the company has plans to expand its installation territory. Lumos offers prefabricated solarscape kits to building contractors nationwide.

Several module manufacturers have partially transparent glass-on-glass modules that will soon be headed toward the US building market. In the years ahead, these PV modules will undoubtedly be used in both prefab and custom residential structures, creating spaces that are as productive as they are attractive.

(Topher Donahue's business, Alpinecreative (www.alpinecreative.com), based in Nederland, Colorado, provides photography and writing for the outdoor recreation and alternative energy industries. The author can be reached at topher@alpinecreative.com).

10

Hybrid Solar Lighting Illuminates Energy Savings for Government Facilities

The article explains that hybrid solar lighting provides new means of reducing energy consumption and delivering significant benefits associated with natural lighting. Hybrid lighting has the potential to significantly reduce energy consumption while also maintaining or exceeding high quality lighting requirements.

Overview

Artificial lighting accounts for the largest component of electricity use in commercial US buildings. Hybrid solar lighting provides an exciting new means of reducing energy consumption while also delivering significant benefits associated with natural lighting in commercial buildings. Hybrid solar lighting contributes to meeting the requirements set by the Energy Policy Act of 2005 for renewable energy consumption by the federal government to be not less than 3% in FY 2007-2009, 5% in FY 2010-2012, and 7.5% in 2013 and thereafter.

The hybrid lighting technology was originally developed for fluorescent lighting applications but recently has been enhanced to work with incandescent accent-

Source: www.ornl.gov/sci/solar © Federal Energy Management Program, US Department of Energy. Reprinted with permission. The article was originally printed in the Department of Energy's (DOE) Energy Efficiency and Renewable Energy's Federal Energy Management Program (FEMP) publication, FEMP Focus.

lighting sources, such as the Parabolic Aluminized Reflector (PAR) lamps commonly used in retail spaces. Commercial building owners—specifically retailers—use the low-efficiency PAR lamps because of their desirable optical properties and positive impact on sales. Yet the use of this inefficient lighting results in some retailers' spending 55-70% of their energy budgets on lighting and lighting-related energy costs.

Hybrid lighting has the potential to significantly reduce energy consumption while also maintaining or exceeding lighting quality requirements. Implementation of the hybrid solar lighting technology across the United States would represent significant energy savings to the country and would provide building managers with a near-term, energy-efficient, higher quality, economically viable alternative to incandescent lamps.

Artificial lighting accounts for almost a quarter of the energy consumed in commercial buildings and 10-20% of energy consumed by industry. Solar lighting can significantly reduce artificial lighting requirements and energy costs in many commercial and industrial buildings and in institutional facilities such as schools, libraries, and hospitals.

Future R&D is aimed at enhancing the performance and reliability of the technology as well as extending the application of the system to work with newly emerging solid-state lighting sources. The hybrid solar lighting technology delivers the benefits of natural lighting with the advantages of an electric lighting system—flexibility, convenience, reliability, and control—and overcomes the constraints that marginalized the use of daylighting in the 20th century.

Principles of Operation

The hybrid solar lighting system uses a roof-mounted solar collector (Figure 1) to concentrate visible sunlight into a bundle of plastic optical fibers. The optical fibers penetrate the roof and distribute the sunlight to multiple "hybrid" luminaires within the building (Figure 2). The hybrid luminaires blend the natural light with artificial light (of variable intensity) to maintain a constant level of room lighting. One collector powers about eight fluorescent hybrid light fixtures, which can illuminate about 1000 square feet.

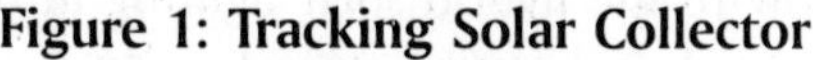

Figure 1: Tracking Solar Collector

Figure 2: Conceptual Illustration of a Hybrid Solar Lighting System

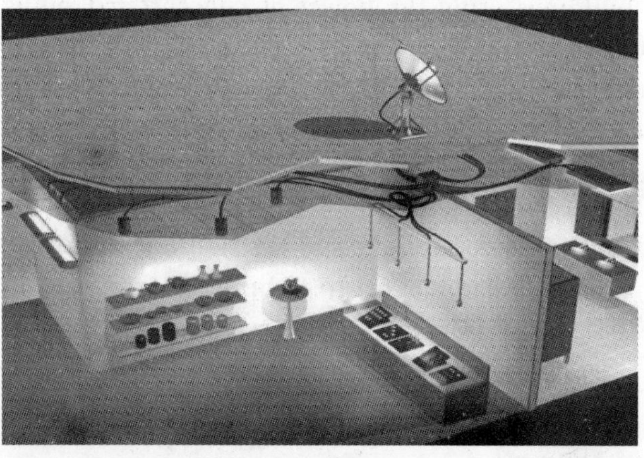

When sunlight is plentiful, the fiber optics in the luminaires provides all or most of the light needed in an area. During times of little or no sunlight, a sensor controls the intensity of the artificial lamps to maintain a desired illumination level. Unlike conventional electric lamps, the natural light produces little to no waste heat, having an efficacy of 200 lumens/Watt (l/W), and is cool to the touch. This is because the system's solar collector removes the infrared (IR) light from the sunlight—the part of the spectrum that generates much of the heat in conventional bulbs. Because the optical fibers lose light as their length increases, it makes sense right now to use hybrid solar lighting in top-story or single-story spaces. The current optimal optical fiber length is 50 feet or less.

The hybrid solar lighting technology can separate and use different portions of sunlight for various applications. Thus, visible light can be used directly for lighting applications while IR light can be used to produce electricity or generate heat for hot water or space heating. The optimal use of these wavelengths is the focus of continued studies and development efforts.

ORNL Studies

Although clean and abundant, solar energy is diffuse and must be captured, concentrated, stored, and/or converted to be used in the highest-value energy form. Under DOE's Office of Energy Efficiency and Renewable Energy, Solar Technologies Program, Oak Ridge National Laboratory (ORNL) has demonstrated the technical feasibility of an entirely new, highly energy-efficient way of lighting and heating buildings using the power of concentrated sunlight. ORNL is currently developing techniques for transporting the sun's energy [visible light and radiant ultraviolet (UV) and IR energy] into buildings for use not only for interior lighting (as shown in Figure 2), but for novel hybrid applications such as

Figure 3: In a Solar Lighting and Power System, the Roof-Mounted Concentrators Collect Sunlight and Distribute it through the Optical Fibers (Enlargement) to Hybrid Lighting Fixtures in the Building's Interior

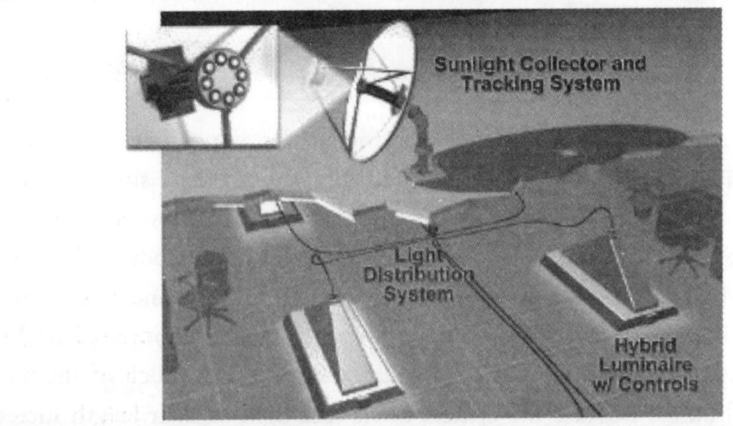

solar hot water heating and space heating. ORNL is also conducting materials and engineering R&D to improve the performance of the tracking mechanisms and fiber optic materials used in these systems.

The goal for future hybrid solar lighting systems is that they be:

- **Multifunctional**—compatible with various electric lamps, light fixtures, hot water heaters, photovoltaics, etc., and usable for various applications
- **Reconfigurable**—easily modified as space needs change
- **Seamlessly integrated**—connected to standard power sources to ensure that disruptions in service do not occur on cloudy days or at night
- **Architecturally compatible**—designed to eliminate architectural design hassles and maintenance problems limiting the use of solar power
- **Affordable.**

The new hybrid solar lighting strategy is the logical pathway for achieving these goals because the technology uses small, flexible optical fibers to deliver the sunlight directly to where it is needed. Hybrid solar lighting is applicable to many building types and can be used for a variety of lighting and heating applications. As a result, the potential near-term energy savings of hybrid solar lighting could be significant.

Advantages of Hybrid Solar Lighting

Electric lighting is the greatest consumer of electricity in commercial buildings (Figure 4), and the generation of this electricity by conventional power plants is the building sector's most significant cause of air pollution. Hybrid lighting can help conserve electricity in proportion to the amount of sunlight available. Hybrid solar lighting technology could benefit federal buildings, particularly in the Sunbelt where cooling is a significant source of energy use.

Full-spectrum solar energy systems provide a new and realistic opportunity for wide-ranging energy, environmental, and economic benefits. Because hybrid solar lighting has no infrared component, it can be considered a high-efficiency light source. Other advantages of hybrid solar lighting are:

Figure 4: Commercial Energy End Uses, from the 2005 Buildings Energy Data Book, US Department of Energy

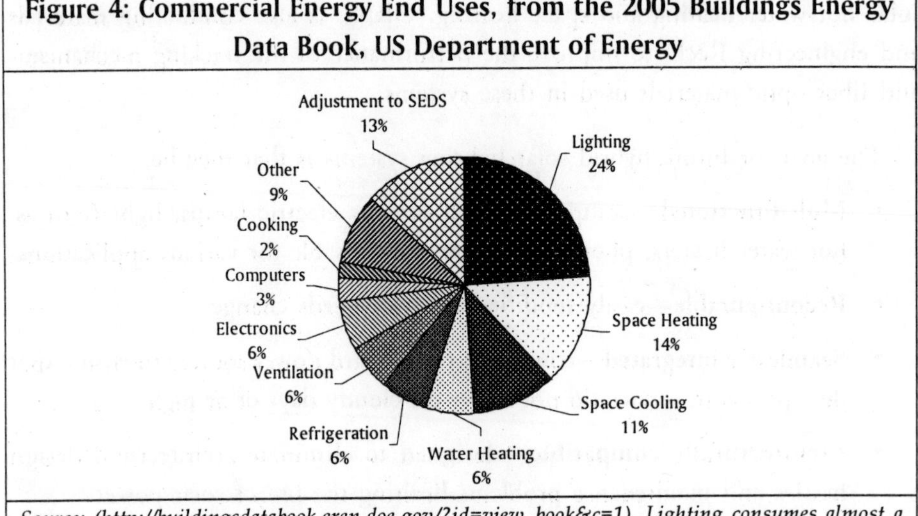

Source: (http://buildingsdatabook.eren.doe.gov/?id=view_book&c=1). Lighting consumes almost a quarter of the electricity used in commercial buildings.

- Roof penetrations are small and minimal, reducing the potential for leaks.
- IR and UV energy in sunlight is separated from the visible light, rather than being transmitted into buildings. Heating, Ventilation, and Air-Conditioning (HVAC) loads are thus reduced by 5 to 10%, compared to buildings having conventional electric lighting systems.
- Hybrid solar lighting systems are readily adaptable to commercial buildings with multiple floors, relatively low ceiling heights, and interior walls, though currently fiber optic output is optimized on the top two floors. A single system can distribute enough sunlight to co-illuminate several rooms in a typical office building.
- Large portions of valuable plenum space—the area between the roof and drop ceiling—are not needed, so there is little competition with other building services, such as HVAC ducts, sprinkler systems, and electrical conduits.
- Hybrid solar lighting can be used both for direct ambient lighting (as in skylights) and for indirect lighting, task lighting, and accent lighting.
- In retrofit applications, hybrid solar lighting is easily incorporated into existing building designs, and the optical fibers can be rerouted to different

locations as lighting needs change. By intentionally misaligning the solar collector from the sun, occupants can even dim or curtail distributed sunlight.

Cost Considerations

The concept of hybrid lighting has existed since the early 1970s, but it has been difficult to make the technology practical. Japanese researchers had earlier developed solar collectors with glass optical fibers—which are more heat-resistant, but also more expensive and harder to work with. The glass-based system costs about $40,000 to illuminate 1000 square feet. Through the use of plastic optical fibers and components, ORNL has been able to cut the system cost significantly, moving closer to a target of $3,000 for a system that illuminates 1000 square feet (see timeline for cost reductions in the following table). When that target price is reached, a building owner in Hawaii could pay for implementing the new technology in just 2-3 years with the savings on electricity bills alone. In other parts of the country, where sunlight is less abundant and utility costs are lower, this payback would obviously take longer.

The payback period for hybrid solar lighting lengthens in proportion to the efficiency of the electric lamps used in combination with distributed sunlight. Because linear fluorescent lamps are very efficient (65-90 lm/W), the models indicate that a hybrid configuration used with such lamps will require more than 10 years to pay for itself in most regions of the country during the early years of commercialization. As prices fall, hybrid solar lighting has the potential to become cost-competitive in most indoor lighting scenarios.

A hybrid configuration is likely to extend the typical life of incandescent and/or halogen lamps. When incandescent lamps are dimmed, filament temperatures decrease. As filament temperatures decrease, life expectancy increases. Although the lamps will last longer, a penalty in efficiency occurs because cooler filaments are generally less efficient at radiating visible light.

In contrast to roof penetrations for skylights, penetrations for hybrid solar lighting are few and small, reducing the potential for leaks.

As R&D improves system performance, increases system lifetime, and reduces system price, and as the secondary benefits of the technology are demonstrated (e.g., improved employee productivity), hybrid solar lighting will move into the larger market of existing buildings that use all fluorescent lighting.

Projected Hybrid Solar Lighting Savings

By the year 2012, hybrid solar lighting should be saving the nation more than 50 million kWh/year while also dramatically improving lighting quality in commercial buildings. Through commercialization efforts with industry partners, more than 5000 hybrid solar lighting systems will be installed by 2012 in regions of the United States where solar availability and electricity rates make this technology cost-effective to consumers. A system tailored to commercial buildings with mixed fluorescent and incandescent lighting (commonly found in retail applications) is likely to be the first application of this technology that will succeed in the marketplace. For this application, a system price of $4000 (installed) has been identified as necessary to produce energy savings of 50 million kWh/year by 2012.

Timeline for Price Reductions in Hybrid Solar Lighting

Cost Element	2006	2007 (product launch)	2012
System cost	$20,000	$16,000	$3,000
Installation cost	$4,000	$3,000	$1,000

Availability

The components of the hybrid solar lighting system are commercially available. ORNL is partnering with industry suppliers of collectors, fiber optic distribution systems, and luminaries (light fixtures) to transfer the technology and make it cost-effective. ORNL maintains patents on the technology and anticipates that the bundled package will be commercially available in 2007.

Potential Candidates for Hybrid Solar Lighting

The first commercial use for hybrid solar lighting will probably be on the upper two floors of buildings having the following characteristics:

(1) Sunbelt location in areas where daytime electricity prices are highest;

(2) occupied every day, including weekends; and

(3) lighting quality (or color rendering) is important and less-efficient electric lamps are currently used.

Hybrid Solar Lighting Technology could Replace Less Efficient Conventional Electric Lamps	
Type of Lighting	**Typical Energy Efficiency (approx. lm/W)**
Incandescent	15
Fluorescent	75
Hybrid Solar	200

Hybrid solar lighting is being targeted first for commercial buildings because in these buildings lighting can account for the largest part of the electricity bill. Residential uses may be farther down the road because the cost advantages there are not as great. However, federal sites would benefit from using hybrid solar lighting technology because decreased use of conventional electric lamps and the air-conditioning loads associated with them would result in energy savings.

Additional benefits would include improved lighting quality and positive environmental impacts from reducing the need to generate electricity. New construction projects create outstanding opportunities for holistic building designs that utilize highly efficient systems that can create satisfying working and living environments with minimized operating costs. Successfully balancing these factors against cost and scheduling issues is the key to creating a well designed and efficiently operating building.

New Developments in Hybrid Solar Lighting Technology

In FY 2005, a fabrication technique was identified for manufacturing an acrylic mirror that meets the tolerances and requirements of a hybrid solar lighting solar collector. A 48-in.-diameter parabolic acrylic mirror is being fabricated by Bennett Mirrors of New Zealand. Early efforts appear promising, with the surface quality of the mirror rivaling that of its glass counterpart. The estimated cost of the new mirror is less than $300 (as compared to $3500 for a glass mirror) and weighs only 9 pounds (as compared to 50 pounds for a glass mirror). Completion of a prototype mirror and manufacture of five mirrors was successful in FY 2006.

The HSL system houses a sun tracking control board that is used to calculate the sun's position based on latitude, longitude and time of day (Figure 5). The control board uses a microprocessor to compute the astronomical equations, obtained by the US Naval Observatory, which are good to 1/60° for the next 300 years. This precision allows the system to track at 0.1° accuracy. The calculations determine the positions in the Azimuth and Zenith Earth-Based coordinate system using latitude, longitude, and Coordinated Universal Time from a global positioning system receiver. The positions are then converted to units of encoder counts for the two encoders used to detect the location of the collector on each axis. The controller compares the actual direction it is pointing to the actual computed position of the sun and then determines if the collector needs to be moved to match its position with that of the sun. The motors then move at a speed proportional to the difference in the actual and computed positions. This process is performed continuously throughout the day in order to track the sun accurately. The control board operates on a 12V or 24V dc supply and uses less than 2 Watts. A photovoltaic solar cell can also be used to power the board.

Figure 5: Controls Enable the Solar Collector to Track the Sun to an Accuracy of One-Tenth Degree

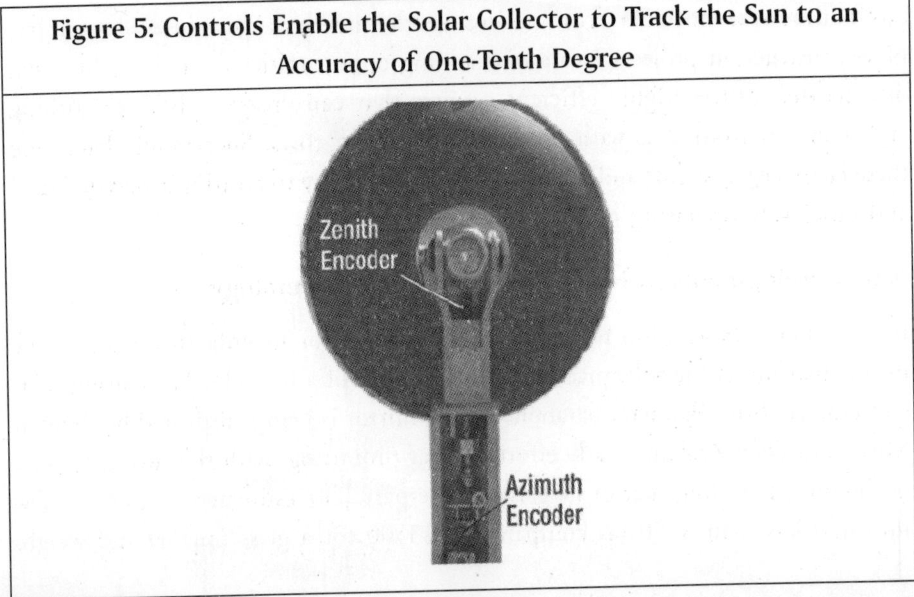

The development of a redesigned and less expensive tracking mechanism for the solar collector was also begun in FY 2005. Among suggested design modifications from an initial "manufacturability" analysis was the use of a high-precision linear actuator in combination with a gear-train drive to reduce cost while still providing high tracking accuracy. The analysis indicated that the tracker cost could be reduced from $25,000 to less than $4000. In 2007, the tracker cost is projected to be reduced to $5250. Detailed modeling and final drawings have been completed, and fabrication of ten new HLS3000 solar tracking units was completed in FY 2006.

Aging of the polymers used in optical fibers to distribute concentrated sunlight is still a concern, as is the need for more efficient methods of coupling converging sunlight into fiber optic bundles. New luminaires that provide seamless spatial and chromatic uniformity during transitions between natural and electric illuminants must be developed for several lighting applications. Ultimately, the advent of hybrid solar and solid-state lighting systems that use light-emitting diodes (LEDs) capable of chromatically adapting to match the spectrum of sunlight throughout the day is expected. An early prototype hybrid LED/hybrid-solar lighting system is shown in Figure 6.

Figure 6: LEDs Now being Developed will Adapt to the Spectrum of Sunlight throughout the Day

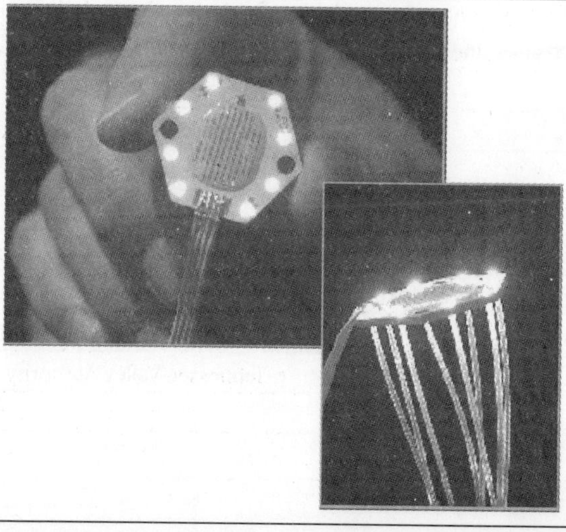

Prototypes are already proving that the hybrid solar lighting concept is viable both technically and economically. The latest prototype provides lighting practitioners with unprecedented design flexibility and control over where and how sunlight is used inside buildings.

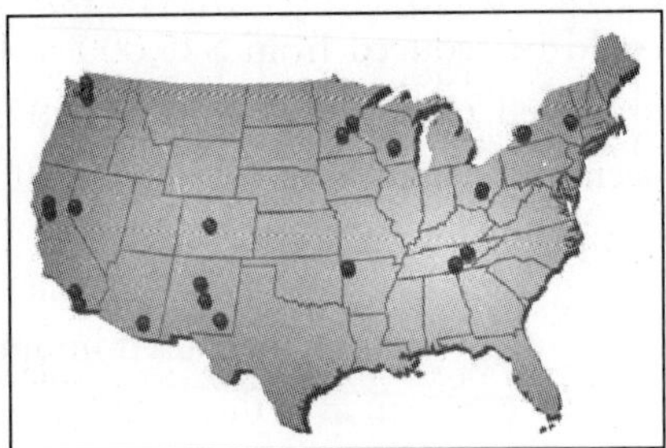

This prototype uses a bundle of 127 small optical fibers, each of which can distribute 350 lumens to several different hybrid luminaires on a sunny day, making possible numerous daylighting applications. For example, some hybrid luminaires being developed allow hybrid solar lighting to be used with linear and compact florescent lamps and with incandescent/halogen lamps.

The Hybrid Lighting Partnership

Industry, Utility, and State Partners	Academic & National Laboratory Partners
• 3M	• Los Alamos National Laboratory
• Advance Transformer Company	• National Renewable Energy Laboratory
• Advanced Lighting Systems, Inc.	• Oak Ridge National Laboratory
• Array Technologies	• Ohio University
• California Energy Commission	• Rensselaer Polytechnic Institute
• Enhancement Electronics, Inc.	• Sandia National Laboratories
• JX Crystals, Inc.	• University of Nevada-Reno
• LSI Industries, Inc.	• University of Wisconsin-Madison
• Protomet	• University of Arizona-Tucson
• ROC Glassworks	
• Science Applications Int'l Corp.	**Partnering Sponsors**
• Sacramento Municipal Utility District	• US DOE (EERE and Fossil Energy)
• Sunlight Direct, Inc.	• California Energy Commission
• The Watt Stopper	• Tennessee Valley Authority
• Tennessee Valley Authority	
• Wal-Mart Stores, Inc.	

Field Trials: The Sunlight Inside Initiative

During a 2006-2007 field trial program at ORNL entitled the "Sunlight Inside Initiative," Sunlight Direct, Inc. (*www.sunlight-direct.com*), working in collaboration with ORNL, is purchasing and installing hybrid lighting system components for each participating host-site. The component suppliers are ready to mass-produce HSL components for Sunlight Direct. The goals of the field trial project are the following:

1. Demonstrate the technical feasibility and energy-savings potential of hybrid solar lighting in the commercial lighting market
2. Expose emerging hybrid solar lighting technology to users and stakeholders across the country
3. Acquire needed experience and field data on hybrid solar lighting equipment and associated installation, operation, and maintenance issues and requirements
4. Create an initial demand for hybrid solar lighting technology by installing 100 or more units in 2008
5. Provide an initial incentive for original equipment manufacturers of hybrid solar lighting to begin investing in automation of component-level manufacturing with the goals of reducing costs and improving market viability.

Sunlight Direct is installing systems at $24,000 per system for ORNL's field trial program, based on individual interest. Built into this cost is a one-year service agreement that covers any maintenance needed on the system. Following the field trial, host sites will be offered an extended service agreement. All installations should be complete by March 2007. ORNL will analyze all data collected from the field trial program and publish the results.

The host sites include the new research facilities on the ORNL campus, the American Museum of Science and Energy (AMSE) in Oak Ridge, the Sacramento Municipal Utility District in Sacramento, California, and a Wal-Mart store in McKinney, Texas. The experience gained from these sites and other field trials will be used to evaluate and improve the technology.

The field trials will establish the hybrid solar lighting technology as a means of saving energy by reducing electrical loads for lighting and by decreasing the amount of heat generated by lighting, thereby decreasing air conditioning costs. This is particularly important in regions where facilities face high peak demand charges on hot summer days. Also, the daylighting character of hybrid solar lighting can greatly improve lighting quality by providing full-spectrum, high-color, low-temperature lighting. S tudies have shown that daylighting can increase productivity in both the workplace and the classroom.

Hybrid Lighting Partnership

A broad-based public-private alliance is working to commercialize hybrid solar lighting. The Hybrid Lighting Partnership includes the organizations listed in the table given in page 140. The partnership's mission is to develop and deploy hybrid solar lighting worldwide early in the 21st century. By 2020 in the United States, hybrid solar lighting is expected to provide:

- annual energy savings of more than 30 billion KWh (> 0.3 Quads),
- reductions in carbon emissions exceeding 5 megatonnes of carbon per year, and
- total economic benefits exceeding $5 billion.

The partnership will improve quality of life by providing more efficient and affordable solar energy, cleaner air, lower utility bills during peak demand periods, and a healthier, more productive work environment.

Future of Hybrid Solar Lighting

Electricity use for artificial lighting in commercial building space costs building owners nearly $17 billion a year (personal communication, Energy Information Administration). Despite the high energy consumption and cost of electric lighting, natural lighting from conventional options such as skylights and windows illuminates only a tiny fraction of existing commercial buildings. This limited use of natural lighting is a result of the architectural limitations of skylights and windows and the uncontrollable nature of the sunlight itself (i.e., it fluctuates in intensity, and it can be highly directional, producing glare and unwanted heating).

The future is bright for hybrid solar lighting. The nationwide field trial program will provide system performance data and user feedback essential for successful use of this solar energy technology. During the field trial program R&D will continue at ORNL in collaboration with industry and university partners to lower component costs, improve the longevity of optic fibers, and advance system control. New solid-state (LED) hybrid luminaires are also being researched for increased energy efficiency. Exciting new areas of R&D for ORNL and its industry and university partners include utilizing hybrid solar lighting technology for space heating, water heating, and hydrogen production.

References

Earl, D, and J. D. Muhs, Oak Ridge National Laboratory, "Preliminary Results on Luminaire Designs for Hybrid Solar Lighting Systems," paper presented at Forum 2001: Solar Energy: The Power to Choose, April 21–25, 2001, Washington, D.C. Online at *http://www.eere.energy.gov/solar/sl_related_links.html.*

Earl, D. D., and L. C. Maxey, "Alignment of an Inexpensive Paraboloidal Concentrator for Hybrid Solar Lighting Applications," SPIE 48th Annual Meeting, San Diego, CA, Political Science and Technology. SPIE, Bellingham, WA. 2003.

Earl, D. D., and R. R. Thomas, "Performance of New Hybrid Solar Lighting Luminaire Design," Proceedings of the 2003 International Solar Energy Conference, Kohala Coast, HI, ASME, JSME, and JSES. ASME International, New York, NY. 2003.

Edwards, L., and P. Torcellini, A Literature Review of the Effects of Natural Light on Building Occupants, NREL/TP-550-30769. National Renewable Energy Laboratory, Golden, CO. July 2002. Online at *http://www.eere.energy.gov/solar/sl_related_links.html.*

Heschong Mahone Group, Inc., "Daylighting and Productivity," *http://www.h-m-g.com/projects/daylighting/projects-PIER.htm.*

Maxey, L. C., D. D. Earl, and J. D. Muhs, "Luminaire Development for Hybrid Solar Lighting Applications," SPIE Proceedings Series, San Diego, CA, International Society for Optical Energy. SPIE, Bellingham, WA. 2003.

Muhs, J. D., Oak Ridge National Laboratory, "Design and Analysis of Hybrid Solar Lighting and Full-Spectrum Solar Energy Systems," paper presented at SOLAR2000 Conference, American Solar Energy Society, June 16–21, 2000, Madison, WS. Online at *http://www.eere.energy.gov/solar/sl_related_links.html.*

Muhs, J. D., Oak Ridge National Laboratory, "Design and Analysis of Hybrid Solar Lighting and Full-Spectrum Solar Energy Systems," paper presented at SOLAR2000 Conference, American Solar Energy Society, June 16-21, 2000, Madison, WS. Online at *http://www.eere.energy.gov/solar/sl_related_links.html.*

Muhs, J. D., Oak Ridge National Laboratory, "Hybrid Solar Lighting Doubles the Efficiency and Affordability of Solar Energy in Commercial Buildings," *CADDET Energy Efficiency Newsletter*, No. 4 (2000), pp. 6-9. Online at *http://www.eere.energy.gov/solar/sl_related_links.html.*

Schlegel, G.O., F. W. Burkholder, S. A. Klein, W. A. Beckman, B. D. Woods, and J. D. Muhs, "Analysis of a Full-Spectrum Hybrid Lighting System," Solar Energy, Volume 76, Issue 4. April 2004.

Ashdown, B.G., D. J. Bjornstad, G. Boudreau, M. V. Lapsa, B. Shumpert, and F. Southworth, *Assessing Consumer Values and Supply-Chain Relationships for Solid State Lighting Technologies*, ORNL/TM-2004/80. Oak Ridge National Laboratory, June 2004.

National Mental Health Association, "Seasonal Affective Disorder" *http://www.nmha.org/infoctr/factsheets/27.cfm.*

11

Solar Greenhouse Resources

Horticulture Resource List

Barbara Bellows and Katherine Adam

This resource list discusses basic principles of solar greenhouse design, as well as different construction material options. Books, articles and Web sites, and computer software relevant to solar greenhouse design are all provided in a resource list.

Introduction

Since 2000, US greenhouse growers have increasingly adopted high tunnels as the preferred solar greenhouse technology. Rigid frames and glazing are still common in parts of Europe, and in the climate-controlled operations in Mexico and the Caribbean that produce acres of winter crops for North American markets. (For more on climate-controlled technology, see Linda Calvin and Roberta Cook. 2005. "Greenhouse tomatoes change the dynamics of the North American fresh tomato industry." *AmberWaves*. April. Vol. 3, No. 2.)

All greenhouses collect solar energy. Solar greenhouses are designed not only to collect solar energy during sunny days but also to store heat for use at night or

Source: www.ncat.org © NCAT. Reprinted with permission. For more information, contact www.ncat.org.

during periods when it is cloudy. They can either stand alone or be attached to houses or barns. A solar greenhouse may be an underground pit, a shed-type structure, or a hoophouse. Large-scale producers use free-standing solar greenhouses, while attached structures are primarily used by home-scale growers.

Passive solar greenhouses are often good choices for small growers because they are a cost-efficient way for farmers to extend the growing season. In colder climates or in areas with long periods of cloudy weather, solar heating may need to be supplemented with a gas or electric heating system to protect plants against extreme cold. Active solar greenhouses use supplemental energy to move solar heated air or water from storage or collection areas to other regions of the greenhouse. Use of solar electric (photovoltaic) heating systems for greenhouses is not cost-effective unless you are producing high-value crops.

Hazards due to increased weather turbulence:

- Hail
- Tornados
- High straight-line winds
- Build-up of snow, ice.

The majority of the books and articles about old-style solar greenhouses were published in the 1970s and 1980s. Since then, much of this material has gone out of print, and some of the publishers are no longer in business. While contact information for companies and organizations listed in these publications is probably out of date, some of the technical information contained in them is still relevant.

The newest form of solar greenhouse, widely adopted by US producers, is high tunnels. The term glazing, as used in this publication, includes reference to polyethylene coverings for hoophouses.

Out-of-print publications often can be found in used bookstores, libraries, and through the inter-library loan program. Some publications are also available on the Internet. Bibliofind is an excellent, searchable Web site where many used and out-of-print books can be located.

As you plan to construct or remodel a solar greenhouse, do not limit your research to books and articles that specifically discuss "solar greenhouses." Since all greenhouses collect solar energy and need to moderate temperature fluctuations for optimal plant growth, much of the information on "standard" greenhouse management is just as relevant to solar greenhouses. Likewise, much information on passive solar heating for homes is also pertinent to passive solar heating for greenhouses. As you look through books and articles on general greenhouse design and construction, you will find information relevant to solar greenhouses in chapters or under topic headings that discuss:

- energy conservation
- glazing materials
- floor heating systems
- insulation materials
- ventilation methods.

In books or articles on passive solar heating in homes or other buildings, you can find useful information on solar greenhouses by looking for chapters or topic headings that examine:

- solar orientation
- heat absorption materials
- heat exchange through "phase-change" or "latent heat storage materials".

Related ATTRA Publications

- Season Extension Techniques for Market Gardeners
- Organic Greenhouse Vegetable Production
- Greenhouse and Hydroponic Vegetable Production Resources on the Internet
- Potting Mixes for Certified Organic Production
- Integrated Pest Management for Greenhouse Crops
- Herbs: Organic Greenhouse Production
- Plug and Transplant Production for Organic Systems
- Compost Heated Greenhouses
- Root Zone Heating for Greenhouse Crops.

This updated resource list includes listings of books, articles, and Web sites that focus specifically on solar greenhouses, as well as on the topics listed above.

Basic Principles of Solar Greenhouse Design

Solar greenhouses differ from conventional greenhouses in the following five ways.(1) Solar greenhouses:

- have glazing oriented to receive maximum solar heat during the winter.
- use heat storing materials to retain solar heat.
- have large amounts of insulation where there is little or no direct sunlight.
- use glazing material and glazing installation methods that minimize heat loss.
- rely primarily on natural ventilation for summer cooling.

Understanding these basic principles of solar greenhouse design will assist you in designing, constructing, and maintaining an energy-efficient structure. You can also use these concepts to help you search for additional information, either on the "Web," within journals, or in books at bookstores and libraries.

Solar Greenhouse Designs

Attached solar greenhouses are lean-to structures that form a room jutting out from a house or barn. These structures provide space for transplants, herbs, or limited quantities of food plants. These structures typically have a passive solar design.

Freestanding solar greenhouses are large enough for the commercial production of ornamentals, vegetables, or herbs. There are two primary designs for freestanding solar greenhouses: the shed-type and the hoophouse. A shed-type solar greenhouse is oriented to have its long axis running from east to west. The south-facing wall is glazed to collect the optimum amount of solar energy, while the north-facing wall is well-insulated to prevent heat loss. This orientation is in contrast to that of a conventional greenhouse, which has its roof running north-south to allow for uniform light distribution on all sides of the plants. To reduce the effects of poor light distribution in an east-west oriented greenhouse, the north wall is covered or painted with reflective material.(2)

Freestanding Shed-Type Solar Greenhouses(2)	
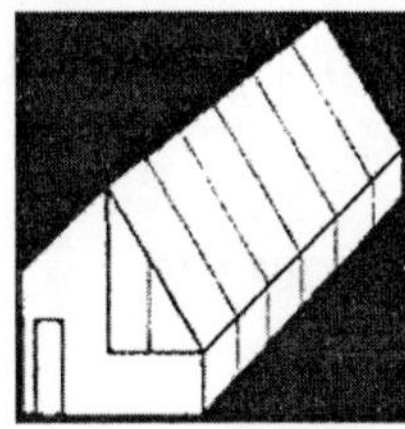	For cold winters, northern latitudes, and year-round use: • steep north roof pitched to the highest summer sun angle for maximum year-round light reflection onto plants. • vertical north wall for stashing heat storage. • 40-60° sloped south roof glazing. • vertical kneewall high enough to accommodate planting beds and snow sliding off roof. • end walls partially glazed for added light. • The Brace Institute design continues the north roof slope down to the ground (eliminating the north wall), allowing for more planting area in ground, but no heat storage against the north wall.
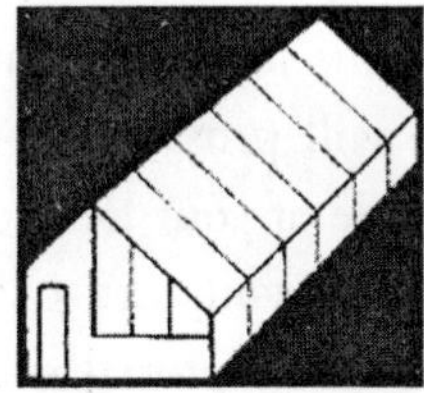	For cold winters, middle US latitudes, and year-round use (similar to the design popularized by Domestic Technology Institute, see Resources for plans and address): • 45-60° north roof slope. • vertical north wall for stacking heat storage. • 45° south roof glazing. • vertical kneewall. • part of end walls glazed for additional light.
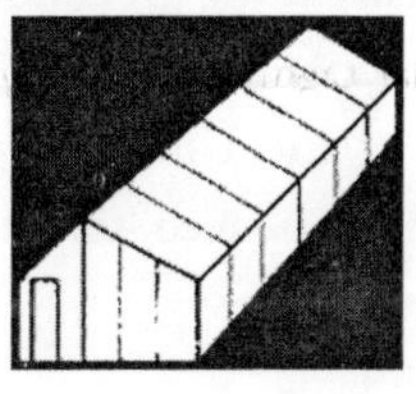	For milder winters, southern US latitudes, and year-round use where less heat storage is needed: • 45-70° north roof slope—roof slope steeper and north wall shorter if less space is needed for stacking heat storage. • roof can extend down to ground, eliminating back kneewall if no storage is used. • 20-40° south roof glazing. • front kneewall as high as is needed for access to beds in front. • most of end walls glazed for additional light.

Freestanding hoophouses are round, symmetrical structures. Unlike the shed-type solar greenhouses, these do not have an insulated north side. Solarization of these structures involves practices that enhance the absorption and distribution of the solar heat entering them. This typically involves the collection of solar heat in the soil beneath the floor, in a process called Earth Thermal Storage (ETS), as well as in other storage materials such as water or rocks. Insulation of the greenhouse wall is important for minimizing heat loss. Heat absorption systems and insulation methods are discussed in detail in the following sections.

Solar Heat Absorption

The two most critical factors affecting the amount of solar heat a greenhouse is able to absorb are:

- The position or location of the greenhouse in relation to the sun
- The type of glazing material used.

Solar Orientation

Since the sun's energy is strongest on the southern side of a building, glazing for solar greenhouses should ideally face true south. However, if trees, mountains, or other buildings block the path of the sun when the greenhouse is in a true south orientation, an orientation within 15° to 20° of true south will provide about 90% of the solar capture of a true south orientation. The latitude of your location and the location of potential obstructions may also require that you adjust the orientation of your greenhouse slightly from true south to obtain optimal solar energy gain.(2) Some growers recommend orienting the greenhouse somewhat to the southeast to get the best solar gain in the spring, especially if the greenhouse is used primarily to grow transplants.(3) To determine the proper orientation for solar buildings in your area, visit the sun chart program at the University of Oregon Solar Radiation Monitoring Laboratory Web page. You need to know your latitude, longitude, and time zone to use this program.

Solar Path at 40° North Latitude

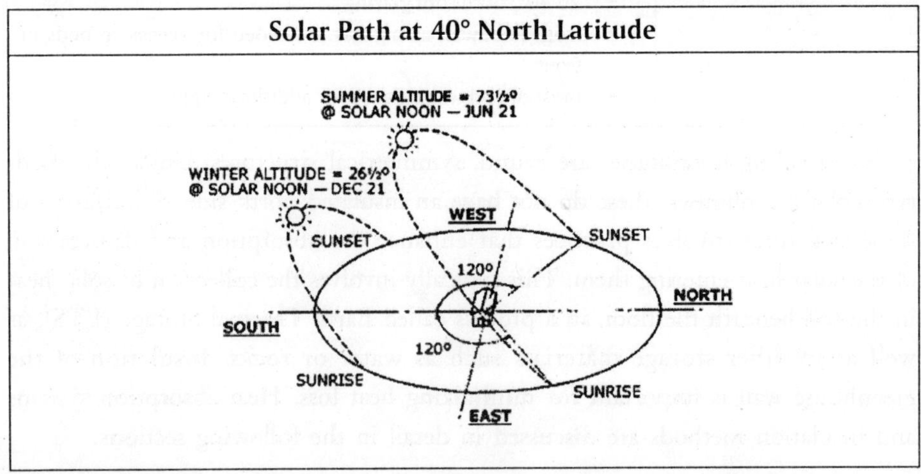

Slope of Glazing Material

In addition to north-south orientation, greenhouse glazing should be properly sloped to absorb the greatest amount of the sun's heat. A good rule of thumb is to add 10° or 15° to the site latitude to get the proper angle. For example, if you are in northern California or central Illinois at latitude 40° north, the glazing should be sloped at a 50° to 55° angle (40° + 10° or 15°).(4)

Glazing

Glazing materials used in solar greenhouses should allow the greatest amount of solar energy to enter into the greenhouse while minimizing energy loss. In addition, good plant growth requires that glazing materials allow a natural spectrum of Photosynthetically Active Radiation (PAR) to enter. Rough-surface glass, double-layer rigid plastic, and fiberglass diffuse light, while clear glass transmits direct light. Although plants grow well with both direct and diffuse light, direct light through glazing subdivided by structural supports causes more shadows and uneven plant growth. Diffuse light passing through glazing evens out the shadows caused by structural supports, resulting in more even plant growth.(5, 6)

Many new greenhouse glazing materials have emerged in recent decades. Plastics now are the dominant type of glazing used in greenhouses, with the weatherability of these materials being enhanced by ultraviolet radiation degradation inhibitors, infrared radiation (IR) absorbency, anti-condensation drip surfaces, and unique radiation transmission properties.(7)

The method used for mounting the glazing material affects the amount of heat loss.(8) For example, cracks or holes caused by the mounting will allow heat to escape, while differences in the width of the air space between the two glazes will affect heat retention. Installation and framing for some glazing materials, such as acrylics, need to account for their expansion and contraction with hot and cold weather.(7) As a general rule, a solar greenhouse should have approximately 0.75 to 1.5 square feet of glazing for each square foot of floor space.(1)

<table>
<tr><th colspan="2">Table 1: Glazing Characteristics</th></tr>
<tr><td>Glass—single layer
Light transmission*: 85-90%
R-value**: 0.9
Advantages:
• Lifespan indefinite if not broken
• Tempered glass is stronger and requires fewer support bars
Disadvantages:
• Fragile, easily broken
• May not withstand weight of snow
• Requires numerous supports
• Clear glass does not diffuse light</td><td>Factory sealed double glass
Light transmission*: 70-75%
R-value**: double layer 1.5–2.0, low-e 2.5
Advantages:
• Lifespan indefinite if not broken
• Can be used in areas with freezing temperatures
Disadvantages:
• Heavy
• Clear glass does not diffuse light
• Difficult to install, requires precise framing</td></tr>
<tr><td>Polyethylene—single layer
Light transmission*: 80-90% - new material
R-value**: single film 0.87
Advantages:
• IR films have treatment to reduce heat loss
• No-drop films are treated to resist condensation
• Treatment with ethyl vinyl acetate results in resistance to cracking in the cold and tearing
• Easy to install, precise framing not required
• Lowest cost glazing material
Disadvantages:
• Easily torn
• Cannot see through
• UV-resistant polyethylene lasts only 1–2 years
• Light transmission decreases over time
• Expand and sag in warm weather, then shrink in cold weather</td><td>Polyethylene—double layer
Light transmission*: 60-80%
R-value** double films: 5ml film 1.5, 6ml, film 1.7
Advantages:
• Heat loss significantly reduced when a blower is used to provide an air space between the two layers
• IR films have treatment to reduce heat loss
• No-drop films are treated to resist condensation
• Treatment with ethyl vinyl acetate results in resistance to cracking in the cold to tearing
• Easy to install, precise framing not required
• Lowest-cost glazing material
Disadvantages:
• Easily torn
• Cannot see through
• UV-resistant polyethylene lasts only 1–2 years
• Light transmission decreases over time
• Expand and sag in warm weather, then shrink in cold weather</td></tr>
<tr><td colspan="2" align="right">Contd...</td></tr>
</table>

Contd...	
Polyethylene—corrugated high density Light transmission*: 70-75% R-value**: 2.5-3.0 *Advantages:* • Mildew, chemical, and water resistant • Does not yellow *Disadvantages:* n/a	**Laminated Acrylic/Polyester film—double layer** Light transmission*: 87%, R-value**: 180% *Advantages:* • Combines weatherability of acrylic with high service temperature of polyester • Can last 10 years or more *Disadvantages:* • Arcrylic glazings expand and contract considerably; framing needs to allow for this change in size • Not fire-resistant
Impact modified acrylic—double layer Light transmission*: 85% *Advantages:* • Not degraded or discolored by UV light • High impact strength, good for locations with hail *Disadvantages:* • Arcrylic glazings expand and contract considerably; framing needs to allow for this change in size • Not fire resistant	**Fiber reinforced plastic (FRP)** Light transmission*: 85-90% - new material R-value**: single layer 0.83 *Advantages:* • The translucent nature of this material diffuses and distributes light evenly • Tedlar-treated panels are resistant to weather, sunlight, and acids • Can last 5 to 20 years *Disadvantages:* • Light transmission decreases over time • Poor weather-resistance • Most flammable of the rigid glazing materials • Insulation ability does not cause snow to melt
Polycarbonate—double wall rigid plastic Light transmission*: 83% R-value**: 6mm 1.6, 8mm 1.7 *Advantages:* • Most fire-resistant of plastic glazing materials	**Polycarbonate film—triple and quad wall rigid plastic** Light transmission*: 75% R-value** triple walls: 8mm 2.0–2.1, 16mm 2.5 R-value** quad wall: 6mm 1.8, 8 mm 2.1 *Advantages:* • Most fire-resistant of plastic glazing materials
	Contd...

Contd...	
• UV-resistant • Very strong • Lightweight • Easy to cut and install • Provides good performance for 7-10 years *Disadvantages:* • Can be expensive • Not clear, translucent	• UV-resistant • Very strong • Lightweight • Easy to cut and install • Provides good performance for 7-10 years *Disadvantages:* • Can be expensive • Not clear, translucent
* *Note that framing decreases the amount of light that can pass through and be available as solar energy* ** *R-Value is a common measure of insulation (hr°Fsq.ft/BTU)*	
Sources: (2, 6, 7, 13, 14)	

You need to understand four numbers when selecting glazing for solar greenhouses. Two numbers describe the heat efficiency of the glazing, and the other two numbers are important for productive plant growth. Many glazing materials include a National Fenestration Rating Council sticker that lists the following factors:

- The SHGC or solar heat gain coefficient is a measure of the amount of sunlight that passes through a glazing material. A number of 0.60 or higher is desired.
- The U-factor is a measure of heat that is lost to the outside through a glazing material. A number of 0.35 BTU/hr-ft2-F or less is desired.
- VT or visible transmittance refers to the amount of visible light that enters through a glazing material. A number of 0.70 or greater is desired.
- PAR or photosynthetically active radiation is the amount of sunlight in the wavelengths critical for photosynthesis and healthy plant growth. PAR wavelength range is 400-700 nanometers (a measure of wavelength).

Note: When choosing glazing, look at the total visual transmittance, not PAR transmittance, to see whether the material allows the spectrum of light necessary for healthy plant growth.

In addition to energy efficiency and light transmission, you should consider the following when choosing glazing materials for your greenhouse:

- Lifespan
- Resistance to damage from hail and rocks
- Ability to support snowload
- Resistance to condensation
- Sheet size and distance required between supports
- Fire-resistance
- Ease of installation

(Based on 6, 9, 10, 11, 12, 13, 14)

Solar Heat Storage

For solar greenhouses to remain warm during cool nights or on cloudy days, solar heat that enters on sunny days must be stored within the greenhouse for later use. The most common method for storing solar energy is to place rocks, concrete, or water in direct line with the sunlight to absorb its heat.(1)

Brick or concrete-filled cinder block walls at the back (north side) of the greenhouse can also provide heat storage. However, only the outer four inches of thickness of this storage material effectively absorbs heat. Medium to dark-colored ceramic tile flooring can also provide some heat storage.(15) Walls not used for heat absorption should be light colored or reflective to direct heat and light back into the greenhouse and to provide a more even distribution of light for the plants.

Storage Materials

The amount of heat storage material required depends on your location. If you live in southern or mid-latitude locations, you will need at least 2 gallons of water or 80 pounds of rocks to store the heat transmitted through each square foot of glazing.(16) If you live in the northern states, you will need 5 gallons or more of water to absorb the heat that enters through each square foot of glazing.(1) Approximately three square feet of four-inch thick brick or cinder block wall is required for each square foot of south-facing glass.(15)

The amount of heat-storage material required also depends on whether you intend to use your solar greenhouse for extending the growing season, or whether you want to grow plants in it year-round. For season extension in cold climates, you will need 2½ gallons of water per square foot of glazing, or about half of what you would need for year-round production.(2)

If you use water as heat-storage material, ordinary 55-gallon drums painted a dark, non-reflective color work well. Smaller containers, such as milk jugs or glass bottles, are more effective than 55-gallon drums in providing heat storage in areas that are frequently cloudy. The smaller container has a higher ratio of surface area, resulting in more rapid absorption of heat when the sun does shine.(14) Unfortunately, plastic containers degrade after two or three years in direct sunlight.

Clear glass containers provide the advantages of capturing heat better than dark metal containers and not degrading, but they can be easily broken.(17)

Trombe walls are an innovative method for heat absorption and storage. These are low walls placed inside the greenhouse near the south-facing windows. They absorb heat on the front (south-facing) side of the wall and then radiate this heat into the greenhouse through the back (north-facing) side of the wall. A Trombe wall consists of an 8- to 16-inch thick masonry wall coated with a dark, heat-absorbing material and faced with a single or double layer of glass placed from 3/4" to 6" away from the masonry wall to create a small airspace. Solar heat passes through the glass and is absorbed by the dark surface. This heat is stored in the wall, where it is conducted slowly inward through the masonry. If you apply a sheet of metal foil or other reflective surface to the outer face of the wall, you can increase solar heat absorption by 30-60% (depending on your climate) while decreasing the potential for heat loss through outward radiation.(10, 18)

Trombe wall

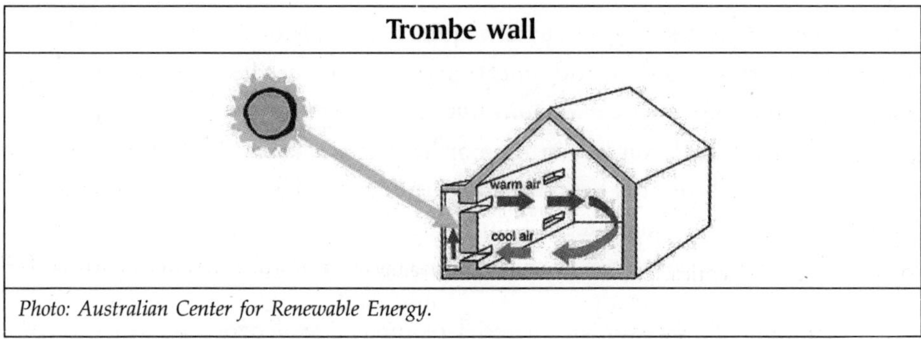

Photo: Australian Center for Renewable Energy.

Water walls are a variation of the Trombe wall. Instead of a masonry wall, water-filled containers are placed in line with the sun's rays between the glazing and the greenhouse working space. The water can be in hard, plastic tubes or other sturdy containers, and the top of the wall can serve as a bench. The Solviva solar greenhouse water wall consists of two 2x4 stud walls, with the studs placed two feet on center. A one-foot spacer connects the two walls. Plastic-covered horse fence wire was then fastened to each stud wall, and heavy-duty, dark-colored plastic water bags were inserted into the space between the two walls. The stud walls were positioned vertically in line with the sun's rays prior to the bags being filled with water.(19) Both the Solviva and Three Sisters Farm Web pages provide designs for constructing solar greenhouses using water walls.

You can use rocks instead of water for heat storage. The rocks should be ½ to 1½ inches in diameter to provide high surface area for heat absorption.(5) They can be piled in wire-mesh cages to keep them contained. Since rocks have a much lower BTU storage value than water (35 BTU/sq.ft/°F for rocks versus 63 for water)(13), you will need three times the volume of rocks to provide the same amount of heat storage. Rocks also have more resistance to air flow than water, resulting in less efficient heat transfer.(20)

Whichever material you choose to use for heat storage, it should be placed where it will collect and absorb the most heat, while losing the least heat to the surrounding air. Do not place the thermal mass so that it touches any exterior walls or glazing, since this will quickly draw the heat away.

Phase-Change

Instead of water or rocks for heat storage, you can use phase-change materials. While phase-change materials are usually more expensive than conventional materials, they are 5 to 14 times more effective at storing heat than water or rocks. Thus, they are useful when space is limited. Phase-change materials include:

- disodium phosphate dodecahydrate
- sodium thiosulfate pentahydrate
- paraffin
- Glauber's salt (sodium sulphate dcahydrate)
- calcium chloride hexahydrate and
- fatty acids (21, 22).

They absorb and store heat when they change from solid to liquid phase, and then release this heat when they change back into a solid phase.(5) Calcium chloride hexahydrate has a heat storing capacity 10 times that of water.(23) These materials are usually contained in sealed tubes, with several tubes required to provide sufficient heat storage. Because of the ability of phase-change materials to absorb high quantities of heat, they also are useful in moderating greenhouse temperatures in the summer.

Most of the research on the use of phase-change materials for greenhouses has been conducted in Europe, Israel, Japan, and Australia. In Israel, phase-change materials were incorporated into greenhouse glazing, which increased heat capture and retention, but reduced the transparency of the glazing on cloudy days when the phase change material did not become liquid.(24) At the time of publication, two companies were identified—one in the US and another in Australia—that sell underfloor heating systems using phase-change materials.(25, 26) Phase-change drywall, currently under research, incorporates phase-change materials inside common wallboard to increase its heat storage capacity and could replace heavier, more expensive, conventional thermal masses used in passive-solar space heating.(27) See the reference section for a listing of publications and Web sites that provide additional information about phase change materials.

For many homeowners, building an attached solar greenhouse is very appealing. They believe that they can extend their garden's growing season while reducing their home heating bills. Unfortunately, there is a contradiction between the use of a greenhouse to grow plants and the use of it as a solar collector for heating the house.(9, 28)

- To provide heat for a home, a solar collector needs to be able to collect heat in excess of what plants can tolerate.
- Much of the heat that enters into a greenhouse is used for evaporating water from the soil and from plant leaves, resulting in little storage of heat for home use.
- A home heat collector should be sealed to minimize the amount of heat loss. Greenhouses, however, require some ventilation to maintain adequate levels of carbon dioxide for plant respiration and to prevent moisture build-up that favors plant diseases.

Bioshelters provide an exception to this rule. In bioshelters, the food producing greenhouse is not an "add-on" to the house but is an integral part of the living space. Bioshelters often integrate fish or small animals with vegetable production to complete nutrient cycles. Biological control measures and plant diversity are used to manage pests in a way that is safe for people and pets in the living quarters. First pioneered by The New Alchemy Institute of East Falmouth, Massachusetts, in the 1970s, Solviva and the Three Sisters Farm carry on the bioshelter tradition.

Active Solar

An active method for solar heating greenhouses uses subterranean heating or earth thermal storage solar heating. This method involves forcing solar-heated air, water, or phase-change materials through pipes buried in the floor. If you use hot air for subsurface heating, inexpensive flexible drainage or sewage piping about 10 centimeters (4 inches) in diameter can be used for the piping. Although more expensive, corrugated drainage tubing provides more effective heating than smooth tubing, since it allows for greater interaction between the heat in the tube and the ground. The surface area of the piping should be equal to the surface area of the floor of the greenhouse. You can roughly calculate the number of feet of four-inch tubing you will need by dividing the square feet of greenhouse floor area by two. Once installed, these pipes should be covered with a porous flooring material that allows for water to enter into the soil around them, since moist soil conducts heat more effectively than dry soil. The system works by drawing hot air collected in the peak of the roof down through pipes and into the buried tubing. The hot air in the tubes warms the soil during the day. At night, cool air from the greenhouse is pumped through the same tubing, causing the warm soil to heat this air, which then heats the greenhouse.(29, 30).

Root-zone thermal heating with water is normally used in conjunction with gas-fired water heaters. This system can be readily adapted to solar and works well with both floor or bench heat. Bench-top heating with root-zone thermal tubing is widely practiced in modern greenhouse production and can be installed easily. A permanent floor heating system consists of a series of parallel PVC pipes embedded on 12" to 16" centers in porous concrete, gravel, or sand. Water is heated in an external solar water heater then pumped into the greenhouse and circulated through the pipes, warming the greenhouse floor. Containerized plants sitting directly on the greenhouse floor receive root-zone heat.

The *Solviva* greenhouse uses a variation of active solar heating. The system in this greenhouse relies on heat absorption by a coil of black polybutylene pipe set inside the peak of the greenhouse. The pipe coil lays on a black background and is exposed to the sun through the glazing. A pump moves water from a water tank, located on the floor of the greenhouse, to the coiled pipe, and back to the tank. Water heated within the coils is capable of heating the water in the tank from 55°F to 100°F on a sunny day. The heat contained in the water tank helps keep the greenhouse warm at night.(19)

Greenhouse management practices also can affect heat storage. For example, a full greenhouse stores heat better than an empty one. However, almost half of the solar energy is used to evaporate water from leaf and soil surfaces and cannot be stored for future use.(5, 31). Besides adding some heat to the greenhouse, increased carbon dioxide in the greenhouse atmosphere, coming from the decomposition activities of the microorganisms in the compost, can increase the efficiency of plant production.

> While solar greenhouses can extend your growing season by providing relatively warm conditions, you should carefully select the types of plants that you intend to grow, unless you are willing to provide backup heating and lighting.
>
> Vegetables and herbs that are suitable for production in a winter solar greenhouse include:
>
> *Cool temperature tolerant*: Basil, celery, dill, fennel, kale, leaf lettuce, marjoram, mustard greens, oregano, parsley, spinach, Swiss chard, turnips, cabbage, collards, garlic, green onions, and leeks.
>
> *Require warmer temperatures*: Cherry tomatoes, large tomatoes, cucumbers (European type), broccoli, edible pod peas, eggplant, and peppers.
>
> (Based on (28))

Insulation

Wall and Floor Insulation

Good insulation helps to retain the solar energy absorbed by thermal mass materials. Keeping heat in requires you to insulate all areas of the greenhouse that are not glazed or used for heat absorption. Seal doors and vents with weather stripping. Install glazing snugly within casements. Polyurethane foams, polystyrene foams, and fiberglass batts are all good insulating materials. But these materials need to be kept dry to function effectively. A vapor barrier of heavy-duty polyethylene film placed between the greenhouse walls and the insulation will keep your greenhouse well insulated.(1) Unglazed areas should be insulated to specifications of your region. For example, R-19 insulation is specified for greenhouses in Illinois (1) and in Missouri (24), while R-21 is recommended for walls in New Mexico.(10) The ZIP-Code Insulation Program Web site provides a free calculator for finding recommended insulation R-values for houses based on your zip code.

Richard Nelson of SOLAROOF developed an innovative way to insulate greenhouse walls in a hoophouse-style greenhouse. This system involves constructing a greenhouse with a double layer of plastic sheeting as glazing. Bubble machines (such as are used to create bubbles at parties) are installed in the peak of the greenhouse between the two layers of plastic. At least two generators should be installed, at either end of the greenhouse. During the winter, the bubble machines face north and blow bubbles into space between two sheets of plastic on the north side of the greenhouse to provide R-20 or higher insulation for northern winters. During the summer, the bubble machines can be turned to face south to provide shading against high heat.(33)

Bubble Greenhouse Design

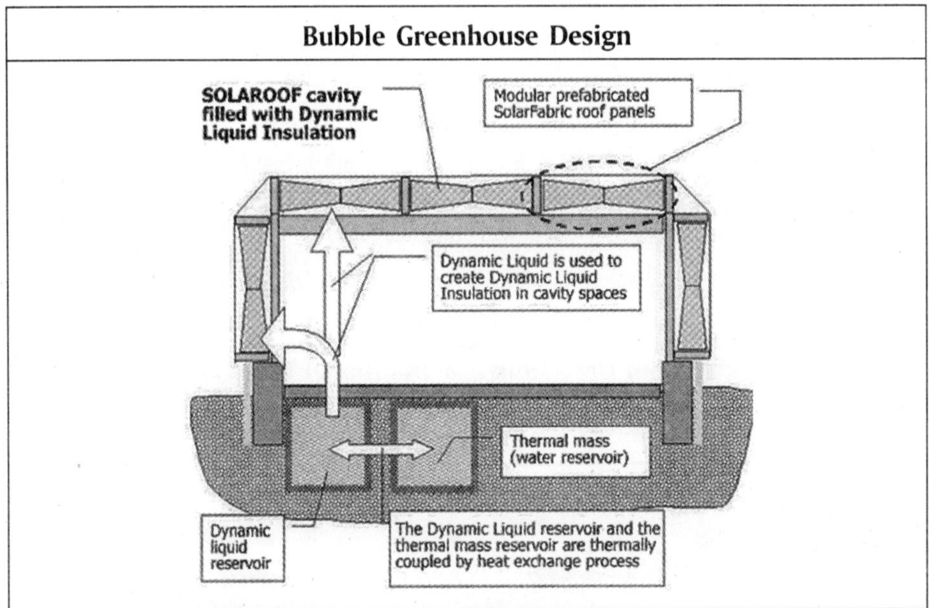

On greenhouse floors, brick, masonry, or flagstone serves as a good heat sink. However, they can quickly lose heat to the ground if there is not an insulating barrier between the flooring and the soil. To protect against heat loss, insulate footings and the foundation with 1- to 2-inch sheets of rigid insulation or with a 4-inch-wide trench filled with pumice stone that extends to the bottom of the footings. You also can insulate flooring with four inches of pumice rock. Besides insulating the floor, this method also allows water to drain through.(16)

External Insulation

You also can insulate your greenhouse by burying part of the base in the ground or building it into the side of a south-facing hill.(5) Straw bales or similar insulating material also can be placed along the unglazed outside walls to reduce heat loss from the greenhouse.(34) Underground or bermed greenhouses provide excellent insulation against both cold winter weather and the heat of summer. They also provide good protection against windy conditions.(35) Potential problems with an underground greenhouse are wet conditions from the water table seeping through the soil on the floor and the entry of surface water through gaps in the walls at the ground level. To minimize the risk of water rising through the floor, build the underground greenhouse in an area where the bottom is at least five feet above the water table. To prevent water from entering the greenhouse from the outside, dig drainage ditches around the greenhouse to direct water away from the walls. Also, seal the walls with waterproof material such as plastic or a fine clay. An excellent description of how to build a simple pit greenhouse is provided at the Web page for the Benson Institute, a division of the College of Biology and Agriculture at Brigham Young University (BYU). This Institute has a campus in Bolivia where students built an underground greenhouse based on local, traditional practices.(36)

The Walipini Greenhouse, a Traditional Underground Greenhouse from Bolivia(36)

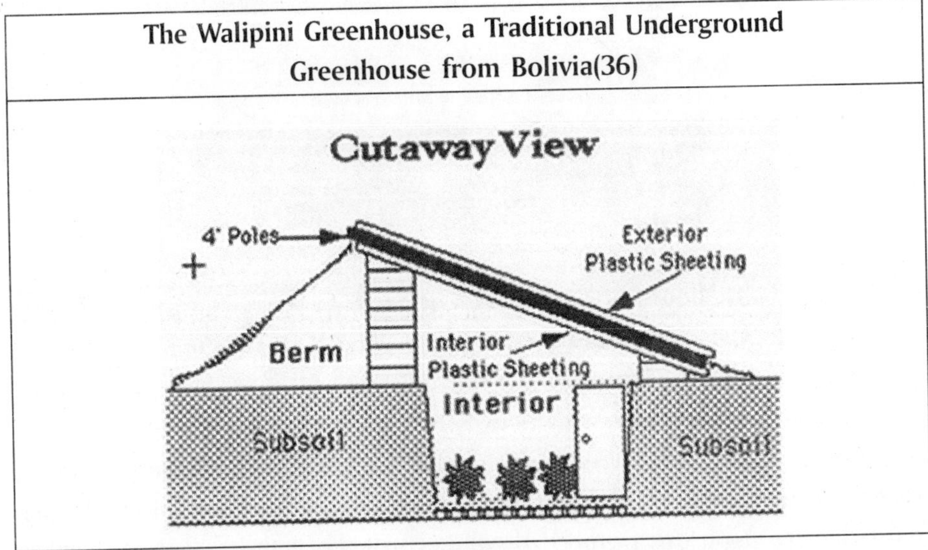

Glazing is what allows light and heat into a solar greenhouse. It can also be the greatest area for heat loss. As mentioned previously, increasing the insulating value of glazing often decreases the amount of sunlight entering the greenhouse. When selecting glazing for your greenhouse, look for materials that provide both good light transmission and insulating value. For example, polyethylene films referred to as "IR films" or "thermal films" have an additive that helps reduce heat loss.(37) Double or triple glazing provides better insulation than single glazing. Some greenhouse growers apply an extra layer of glazing—usually a type of film—to the interior of their greenhouses in winter to provide an extra degree of insulation. Adding a single or double layer of polyethylene film over a glass house can reduce heat loss by as much as 50%.(38) By using two layers of polyethylene film in plastic-film greenhouses with a small fan blowing air between them to provide an insulating air layer, heat losses can be reduced by 40% or more, as compared to a single layer of plastic.(39)

Solar Greenhouse with Solar Curtains, Water Wall, and Water Heat Storage on the North Wall(2)

Greenhouse curtains limit the amount of heat lost through greenhouse glazing during the night and on cloudy days. By installing greenhouse insulation sheets made from two-inch thick bats of polystyrene, you can reduce by almost 90% the heat that would otherwise be lost through the glazing. For a small greenhouse where labor is not a large constraint, you can manually install the polystyrene sheets at night and remove them in the morning. Magnetic clips or Velcro fasteners will facilitate the installation.(1) Alternatively, you can install thermal blankets made of polyethylene film, foam-backed fiberglass, or foil-faced polyethylene bubble material. These blankets are supported on wire tracks and can be raised or lowered using pulleys. While greenhouse curtains composed of thermal blankets are usually opened and closed manually, a few manufactures have motorized roll-up systems that store the blanket near the greenhouse peak.(5)

Ventilation

A building designed to collect heat when temperatures are cold also needs to be able to vent heat when temperatures are warm. Air exchange also is critical in providing plants with adequate levels of carbon dioxide and controlling humidity. Because of the concentrated air use by plants, greenhouses require approximately two air exchanges per minute (in contrast to the one-half air exchange per minute recommended for homes). To determine the flow requirements for your greenhouse, multiply the volume of the greenhouse by two to get cubic feet of air exchange per minute, which is the rate used in determining the capacity of commercial evaporative coolers.

Roof-ridge and sidewall vents provide natural ventilation. The sidewall vents allow cool air to flow into the sides of the greenhouse, while ridge vents allow the rising hot air to escape. Some wind is necessary for this type of ventilation system to function effectively. On still, windless days, fans are necessary to move air through the greenhouse. The area of the venting should be equal to between 1/5 to 1/6 of the greenhouse floor area.(1)

A Solar Chimney(2)

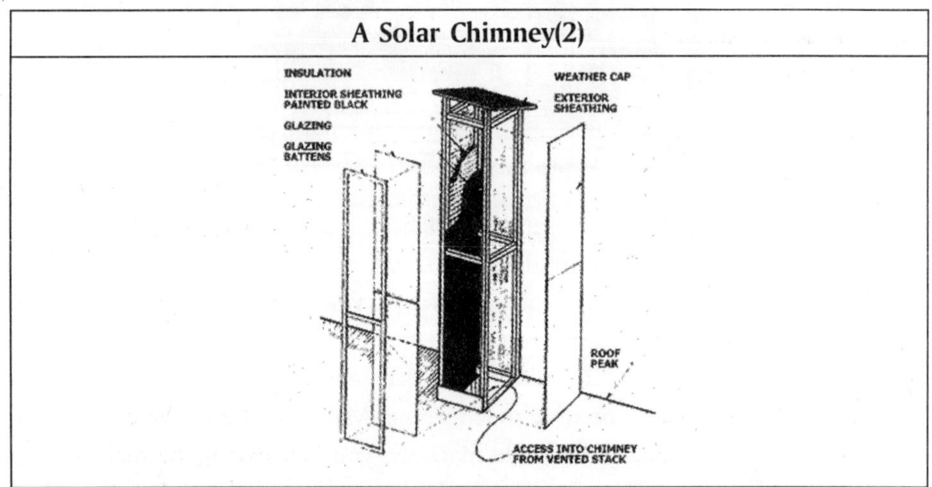

Solar chimneys are passive solar collectors attached to the highest point on the greenhouse and are combined with vents or openings on either end of the greenhouse. The chimney has an inlet that draws warm air from inside the greenhouse and an outlet that discharges it to the outdoors. To enhance solar gain inside the chimney and increase airflow, the inner surface of the chimney stack is glazed or painted black. A ventilator turbine added to the top of the chimney provides an additional force to pull warm air up from inside the greenhouse.(40)

Thermal storage materials are effective in keeping a greenhouse cool in summer as well as keeping it warm in winter. Since these materials absorb heat during the day, less heat radiates within the greenhouse when the sun is shining. When the sun goes down, heat released from the thermal storage materials can be vented out of the greenhouse.(2)

Removing external shading can also decrease heat build-up within the greenhouse. Shading provided by mature trees is not recommended. Older books on solar greenhouse design (e.g., 2) argue that deciduous trees can provide shade in the summer but allow for plenty of sunlight to enter through the glazing in the winter after the leaves are gone. However, more recent literature notes that a mature, well-formed deciduous tree will screen more than 40% of the winter sunlight passing through its branches, even when it has no leaves.(31)

Active solar cooling systems include solar air-conditioning units and photovoltaics set up to run standard evaporative cooling pads. Both are more complex and expensive to equip than passive systems.

Putting it All Together

Designing and building a solar greenhouse can be an exciting and rewarding project. Feel free to rely on the older literature to provide you with basic siting, design, and construction guidelines. However, incorporating new glazing, heat storage, and insulating materials into your design can greatly enhance the efficiency of your structure. Several consulting companies can provide you with blueprints and design assistance, often at a reasonable cost. Of course, you need to weigh the costs of these new technologies against the value of your greenhouse-grown crops. As you become familiar with the principles of passive solar design, you may want to experiment with ways of harnessing the power of the sun within your greenhouse to produce better plants throughout the year.

(Barbara Bellows is highly knowledgeable environmental scientist and collaborative project manager with over 20 years of experience. She possesses excellent writing skills and is author of over 25 published papers and technical manuals.

Katherine Adam (Program Specialist) can be reached at kadam@ncat.org).

References

1. Illinois Solar Energy Association. 2002. Solar Greenhouse. ISEA Fact Sheet #9. Accessed at: *www.illinoissolar.org/*
2. Alward, Ron, and Andy Shapiro. 1981. Low-Cost Passive Solar Greenhouses. National Center for Appropriate Technology, Butte, MT. 173 p.
3. White, Joe. 1991. Growing it in a Sunpit. The Natural Farmer. Winter. p. 14.
4. Thomas, Stephen G., John R. McBride, James E. Masker, and Keith Kemble. 1984. Solar Greenhouses and Sunspaces: Lessons Learned. National Center for Appropriate Technology. Butte, MT. 36 p.
5. Bartok, Jr., John W. 2000. Greenhouses for Homeowners and Gardeners. NRAES-137. Cornell University, Ithaca, NY. 214 p.
6. Giacomelli, Gene A. 1999. Greenhouse coversing systems—User considerations. Cook College. Rutgers University. Accessed at: *http://AESOP.RUTGERS.EDU/~ccea/publications.html*
7. Giacomelli, Gene A. 1999. Greenhouse glazings: Alternatives under the sun. Department of Bioresource Engineering. Cook College. Rutgers University. Accessed at: *http://AESOP.RUTGERS.EDU/~ccea/publications.html*
8. Bartok, Jr., John W. 2001. Energy Conservation for Commercial Greenhouses. NRAES-3. Cornell University, Ithaca, NY. 84 p.
9. BTS. 2001. Passive Solar Design. Technology Fact Sheet. US Department of Energy. Office of Building Technology, State and Community Programs. Accessed at: *www.eere.energy.gov/buildings/info/documents/pdfs/29236.pdf* (PDF/232 K)
10. Luce, Ben. 2001. Passive Solar Design Guidelines for Northern New Mexico. New Mexico Solar Energy Association. Accessed at: *www.nmsea.org/Curriculum/Courses/Passive_Solar_Design/Guidelines/Guidelines.htm*
11. NREL. 2001. Passive Solar Design for the Home. Energy Efficiency and Renewable Energy Clearinghouse. National Renewable Energy Laboratory. US Department of Energy. Accessed at: *www.nrel.gov/docs/fy01osti/27954.pdf* (PDF/216 K)

12. BTS. 2001. Passive Solar Design. Technology Fact Sheet. US Department of Energy. Office of Building Technology, State and Community Programs. Accessed at: *www.nrel.gov/docs/fy01osti/29236.pdf* (PDF/232 K)
13. Smith, Shane. 2000. Greenhouse Gardener's Companion: Growing Food and Flowers in Your Greenhouse or Sunspace. Fulcrum Publishers. 2nd edition. 544 pages. Excerpts accessed at: *www.greenhousegarden.com/energy.htm*
14. Nuess, Mike. 1997. Designing and building a solar greenhouse or sunspace. Washington State University Energy Program.
15. Williams, Sue E., Kenneth P. Larson, and Mildred K. Autrey. 1999. Sunspaces and Solar Porches. The Energy Event. Oklahoma State Cooperative Extension Service. A hard copy can be purchased via the following website *www.osuums.com/ASPFiles/inventfind.asp?s=*.
16. Anon. n.d. Solar Greenhouse Plans and Information. Sun Country Greenhouse Company. Accessed at: *www.hobby-greenhouse.com/FreeSolar.html*
17. North Carolina Solar Center. 2000. Do It Yourself Solar Applications: For Water and Space Heating. North Carolina Solar Center. Energy Division North Carolina Department of Commerce. Accessed at: *www.ncsc.ncsu.edu/information_resources/factsheets/23lowcst.pdf* (PDF/713 K)
18. NREL. 1999. Building a Better Trombe Wall. National Renewable Energy Laboratory.
19. Edey, Anna. 1998. Solviva: How to Grow $500,000 on One Acre and Peace on Earth. Trailblazer Press, Vineyard Haven, MA. 225 p.
20. Pin, Nick. 1995. Solar closets in a nutshell. Listserv message. Archived at: *www.ibiblio.org/london/renewable-energy/solar/Nick.Pine/msg00026.html*
21. Solar Technologies. Accessed at: *www.alaskasun.org/pdf/SolarTechnologies.pdf* (PDF/328 K)
22. Gates, Jonathan. 2000. Phase Change Material Research. Accessed at: *http://freespace.virgin.net/m.eckert/index.htm*
23. Baird, Stuart, and Douglas Hayhoe. 1983. Passive Solar Energy. Energy Fact Sheet.
24. Korin, E., A. Roy, D. Wolf, D. Pasternak, and E. Rappaport. 1987. A novel passive solar greenhouse based on phase-change materials. International Journal of Solar Energy. Volume 5. p. 201–212.
25. PCM Thermal Solutions. Underfloor heating. Accessed at: *www.pcm-solutions.com/under_app.html*
26. TEAP Energy. 2002. PCM Energy Efficiency.
27. EREC. n. d. Phase Change Drywall. EREC Reference Briefs. US Department of Energy. Office of Energy Efficiency and Renewable Energy. Accessed at: *www.eere.energy.gov/troughnet/pdfs/tamme_concrete_tes.pdf*

28. Butler, Nancy J. 1985. A Home Greenhouse—Dream or Nightmare? Weed 'Em and Reap; Feb.-March. MSU Cooperative Extension Service. Accessed at: *www.hobby-greenhouse.com/UMreport.htm*

29. Monk, G.J., D.H. Thomas, J.M. Molnar, and L.M. Staley. 1987. Solar Greenhouses for Commercial Growers. Publication 1816. Agriculture Canada. Ottawa, Canada.

30. Puri, V.M., and C.A. Suritz. 1985. Feasibility of subsurface latent heat storage for plant root zone and greenhouse heating. American Society of Agricultural Engineers (Microfiche collection) 20 p.

31. NREL. 1994. Sunspace Basics. Energy Efficiency and Renewable Energy Clearinghouse. National Renewable Energy Laboratory. US Department of Energy. Accessed at: *www1.eere.energy.gov/office_eere/pdfs/solar_fs.pdf* (PDF/220K)

32. Thomas, Andrew L., and Richard J. Crawford, Jr. 2001. Performance of an Energy-efficient, Solar-heated Greenhouse in Southwest Missouri. Missiouri Agricultural Experiment Station. Missouri University College of Agriculture, Food, and Natural Resources.

33. Nelson, Richard. Sola Roof Garden. Accessed at: ***http://solaroof.org/wiki/SolaRoof/SolaRoofGarden/***

34. Cruickshank, John. 2002. Solar Heated Greenhouses with SHCS. Growing Concerns. Accessed at: ***www.sunnyjohn.com/indexpages/shcs_greenhouses.htm***

35. Geery, Daniel. 1982. Solar Greenhouses: Underground. TAB Books, Inc. Blue Ridge Summit, PA. 400 p.

36. Benson Institute. n.d.. The Pankar-huyu and Building a Pankar-huyu. Accessed at: ***http://benson.byu.edu/Publication/BI/Lessons/volume22/pankar.html*** and ***http://benson.byu.edu/Publication/BI/Lessons/volume22/building.html***

37. Anon. 2002. Greenhouse Glazing. Horticultural Engineering, Rutgers Cooperative Extension, Volume 17, No. 1. Accessed at: ***http://www.rosesinc.org/ICFG/Join_ICFG/2002%20-%2003/Greenhouse_Glazing.asp***

38. Aldrich, Robert A., and John W. Bartok, Jr. 1989. Greenhouse Engineering. NRAES-33. Northeast Regional Agricultural Engineering Service, Cornell University. 203 p.

39. Hunt, John N. 1988. Saving energy—North Carolina style. Greenhouse Grower. March.

40. Gilman, Steve. 1991. Solar ventilation at Ruckytucks Farm. The Natural Farmer. Winter. p. 15.

Section III

Experiences and Initiatives

12

Electricity from Solar Home Systems in South Africa

Gisela Prasad

Developing countries like South Africa have an economic agenda of access to electricity for the poor in remote and rural areas. This paper describes solar water heaters and electricity from solar home systems, both of which include the impact of poverty on the dissemination and acceptance of technology.

1. Introduction

In developed countries, Renewable Energy (RE) technologies are most often introduced for environmental reasons, to reduce GHG emissions mandated under the Kyoto Protocol – which South Africa signed in 2002. The Protocol does not commit non-Annex 1 (developing) countries such as South Africa to any emission targets in the first commitment period (2008 to 2012), however, and it creates no external pressure to reduce emissions. So it is understandable that in this case study the major government concern is not the environment, but access to electricity for the poor in remote rural areas.

Source: www.erc.uct.ac.za © Gisela Prasad. Reprinted with permission.

RE technologies are not widely disseminated in South Africa, although solar resources are very high and solar technologies are particularly suitable. The general environmental awareness is limited when compared to European countries and it is only recently that the media have been more regularly covering issues such as global warming and its impact on South Africa.

The South African government generally supports RE, and its RE policy stipulates a voluntary target of 10,000 GWh to be supplied from renewable sources by 2013. The target is approximately 10% of the country's electricity demand, of which now less than 1% is met from renewable sources (DME 2004). Different players in projects and the industry give various explanations and reasons why the market has not responded more positively, often citing high initial capital cost as the major explanation.

The two South African case studies describe solar water heaters (SWHs) (case study 1) and, in this report, electricity from solar home systems (case study 2). Both case studies include the impact of poverty on the dissemination and acceptance of the technology.

SHS using photovoltaic panels to generate electricity have been provided as part of the National Electrification programme in remote poor rural areas to which the grid has not been extended, as a substitute for grid electricity, although in fact subsidised SHS were expected to bring light and television services at a much faster rate than they actually did.

2. Country Overview

South Africa, like other transition countries, faces the dual challenge of pursuing economic growth and environmental protection. Sustainable energy systems, based on RE resources, offer an opportunity to protect the environment and create economic growth. The implementation of RE technologies faces a major challenge because South Africa has very large coal deposits and the electricity generated from it is amongst the cheapest in the world. The powerful national electricity company Eskom is government-owned and has almost a monopoly of electricity generation; generation is largely by municipalities.

South African energy policies have always been linked to the prevailing political situation. Predemocracy energy policy and planning were characterised by energy security concerns and racially skewed provision. After 1994 the new democratic government addressed the inequalities of the past and electrification of previously disadvantaged populations a priority area identified in the National Reconstruction and Development Programme (RDP). The highly subsidised National Electrification Programme (NEP) increased electricity coverage from about 36% in 1994 to over 70% in 2002.

Even after being connected to the national grid, many poor households could not use the electricity because they were not able to afford it, and continued to cook with kerosene and wood. The electricity consumption rate among the poor therefore remained extremely low. When government realised that the poor did not fully benefit from the large investment in electrification the Free Basic Electricity Policy was introduced, in 2004. Those connected to the grid now receive 50 kWh free every month, sufficient for lighting, black-and-white television, radio and occasional basic cooking. The government pays this subsidy to the municipalities.

Poor rural households have least access to electricity, and providing it to them is a great challenge. Extending the grid to every household in the country is not feasible now for technical and financial reasons, and the question arises as to what distance from the grid makes decentralised electricity supply, such as PV systems for each home, the most appropriate solution – even if the grid is gradually expanded. The SHS programme was designed to give more rural people access to limited electricity until such a time that they get grid connections. Solar cells are imported and some of the systems are assembled in the country. The extent to which NGOs and technology providers pushed the programme has not yet been explored.

As part of the NEP, solar electrification projects were implemented in some of the more remote rural areas. As with grid electrification, government heavily subsidised this programme, with recipients of SHS paying about R120[1], a fraction of the actual cost of approximately R3500 for the system. The service provider owns the SHS and charges a monthly fee of R58 for service and maintenance.

1 1 € is equivalent to R9.30 (April 2007).

Renewable energy is one of the areas the government pursues in managing energy-related environmental impacts and diversifying energy supplies from a coal-dominated system. As mentioned, there are no external pressures on South Africa to reduce GHG emissions and to disseminate RE technologies. The government's White Paper on Renewable Energy Policy (2004) supports the establishment of RE technologies, targeting the provision of 10,000 GWh of electricity from renewable resources by 2013. This has the potential to create 35,000 jobs, adding R5 billion to the GDP and R687 million to the incomes of low-income households (DME, 2004). Solar water heating and biodiesel have the greatest potential to contribute to meeting the target. RE is to be utilised for both power generation and non-electric technologies such as solar water heating and biofuels. By late 2005 the Department of Minerals and Energy (DME) completed a Renewable Energy Target Monitoring Framework to ensure that progress towards the 2013 target is effectively monitored (DME 2005).

South Africa experiences high levels of solar radiation, with average daily solar radiation of between 4.5 kWh and 6.5 kWh per square metre. This resource is relatively predictable and well distributed throughout the country with some regional variations.

The provision of hot water using solar technologies has the benefit of saving households money over the long term and mitigating GHG emissions associated with fossil fuel usage. SWHs are also the least expensive means of heating water for domestic use on a life cycle cost basis because solar energy is free (Austin & Morris 2005).

3. Case Study: Solar Home Systems in South Africa

South Africa is committed to provide universal access to electricity by 2012 (Mlambo-Ngcuka, 2004). Grid electricity is the general approach and about 70% of households are already connected. For the remaining households the Energy White Paper indicates that government will determine an appropriate mix between grid and non-grid technologies, and 'in remote rural areas where the lowest capacity grid system cannot be supplied within the capital expenditure limit, this situation will provide a natural opportunity for Remote Area Power Supply (RAPS) systems to be supplied' (DME 1998). In 1999, about 51% of rural households were still without electricity and it became clear that the supply technology had to be re-evaluated. Photovoltaic SHS were selected to provide a basic service to those households that cannot be grid-connected within acceptable cost parameters (Kotze 2000).

Figure 1: Solar Home System Mounted on a Pole Next to the House, as Roof Structures are Often not Suitable for Supporting the System

Figure 2: Indoor Light from Solar Home System

A pure commercial model and a utility model were considered for supplying SHS to rural households and, innovative in the South African context, it was decided to select the utility model and to involve the private sector (Kotze 1997; 1998). The programme grants private companies the rights to establish off-grid energy utilities in designated concession areas. This utility service provision is a fee-for-service model including the maintenance of the off-grid energy systems by the utility, which has exclusive rights to government subsidies to cover most of the capital costs for five years. The fee-for-service agreement will last for 20 years (Afrane-Okese & Thom 2001).

It was clear from the beginning that the poor rural households for which the systems were intended would not be able to afford the initial capital cost, and a government subsidy of R3500 for each installed system was included in the programme for the first five years. The subsidy was paid directly to the service provider. The customer had to pay R110 as an installation fee and a cellular phone charger was offered for an additional R20.

In 2004, the government introduced a subsidy for free basic electricity for grid-connected households, equivalent to 50 kWh per month. SHS users in the concession areas received an equivalent monthly subsidy of R40, reducing the fee charged for maintaining and servicing the system to R18 per month for each household.

It is still doubtful if very poor rural people can afford even this highly subsidised service of PV just for lighting and media use. A survey of 348 households in the Eastern Cape Province compared access to electricity of both off-grid and grid and income and found that the poorest households (average monthly income R819, equivalent to about €90) remained without any supply of electricity. Only households in the highest income group (R2307 per month) could afford solar electricity, and grid users in neighbouring areas where grid electricity was provided

Table 1: Concessionaires, Concession Areas and Total Number of Installations, June 2004

Concessionaire	Concession Area	Total Number of Installations
Nuon-Raps (NuRa)	Northern Kwa-Zulu Natal	6541
Solar Vision	Northern Limpopo	4758
Shell-Eskom Replaced by 3 smaller companies in 2005/6	Northern parts of the Eastern Cape and Southern Kwa-Zulu Natal	5800
EDF-Total (KES)	Interior Kwa-Zulu Natal	3300
Renewable Energy Africa (REA)	Central Eastern Cape	0
Total		20 399

Source: Willemse (2004); ERC (2004).

had an average income of R1860 (ERC 2004). There is also the question of whether and for how long the government will feel it can afford the high capital subsidy for each system.

3.1 Step One: Vision of the Solar Home System Project

South Africa's high solar radiation means that the PV technology to generate electricity can be used almost anywhere in the country. PV technology is modular, allowing for upscaling or downscaling. PV systems of various sizes can meet a range of electricity needs but are not economic for thermal applications. The government's vision to supply photovoltaic SHS through the private utility model was as follows (Kotze 2000):

- It would speed up universal access to electricity as envisioned in the Energy White Paper since non-grid electricity service had become increasingly cost-effective in remote areas.
- It could attract larger, better organised private companies with their own sources of financing.
- It would facilitate and rationalise electrification planning, funding and subsidisation at national level, allowing regulation and financing mechanisms to maximise targets and optimise resource allocation.
- It had the potential to reduce equipment costs (through volume discounts), transaction costs, and operation and maintenance costs (through economies of scale).
- It ensures service to customer over a long period of time (e.g. 20 years).
- The utility would own the hardware as assets, which should facilitate the raising of capital on the money markets, while the strong financial and maintenance controls characteristic of the private sector should facilitate the channeling of international development funding.
- This should facilitate relocation of technologies that may arise over time as the grid reaches more remote areas.
- It was expected that the service providers would adopt a delivery model that promotes a range of fuels such as gas or kerosene, in addition to SHS

or mini-grid systems. This energisation model has been motivated by the realisation that electricity often does not meet all the energy needs of rural people who, after electrification, tend to continue to rely on multiple fuels.

- Most rural dwellers that have access to grid electricity are usually not able to afford higher consumption of electricity and they tend to use it mainly for lighting, radio and monochrome television, services that can be equally provided by SHS. The service level that is subsidised under the non-grid electrification programme was set at 50 Wp.

The main disadvantages of the utility route were considered to be that the systems were installed at the clients' premises under their control but not under their ownership since the utility owned the systems and they were therefore prone to vandalism, neglect and misuse.

The service level of the subsidized SHS is limited and SHS technology is not very flexible and is limited in its application. The major energy requirement of poor households is cooking, for which PV systems do not provide energy, and higher-power media appliances such as colour televisions usually require a larger PV system than the standard 50 Wp SHS, as does refrigeration.

Other weak points of the SHS utility model are that the systems are expensive, requiring large subsidies in order to be affordable for the rural households and a reasonable commercial venture for the supply utilities. Maintenance in very remote rural areas with poor roads can be problematic. The payment of regular monthly service fees is difficult for households with low and irregular incomes. (In one of the concessions, the utility provided SHS to only those households with proof of regular income, effectively excluding the poor.)

SHS, the concession approach and the fee-for-service model are replicable in any rural area without grid electricity supply. A basic maintenance service is required and the battery has to be replaced at least every 3 to 4 years.

Solar concessions are not financially viable without the capital subsidy for new installations and the operational subsidy. The government seems to be deciding the replicability question by limiting the funds available for capital-cost subsidy and in 2005 government stopped paying the subsidy altogether. The future payment of the monthly operational subsidy is also doubtful. Unless something changes, the whole SHS programme may slowly come to an end.

3.2 Step Two: What were the Various Expectations of the Case?

What Types of Interests/Actors became Involved in Renewable Energy Initiatives at the Level of the Case?

The off-grid concession approach is being tried in four quite remote rural areas, chosen in relation to the national grid and in four provinces (Eastern Cape, KwaZulu-Natal, Mpumalanga and Limpopo) in areas where it is unlikely that the grid will soon reach. However, some households, which had opted for a SHS have recently been connected to the grid, suggesting that electrification plans have either changed or have not been clearly communicated to the SHS providers.

The Eskom-Shell Joint Venture in the Eastern Cape was the first concessionaire to install SHS and others followed, learning from their experience.

The major stakeholders directly involved in the programme are the customers and the service providers. Eskom and municipalities are the licensed electricity providers and they have to demarcate areas in their licence area in which the off-grid service providers can operate and where grid electricity is not going to be provided in the near future. Transparent electrification planning is necessary and should be communicated to the SHS service providers. The Department of Minerals and Energy is to facilitate the process, formulate policy and administer the capital subsidy for the systems and their installation. The Department of Provincial and Local Government is charged with providing services and channeling the free basic electricity subsidy to the service providers. The Electricity Regulator approves the installation of the systems according to the set standards. Service providers are paid the capital subsidy only after the Regulator has approved the installation. The commercial providers of PV systems sell, install and manufacture components.

There are high capacity development needs in the villages where SHS are installed. Training local technicians to do O&M services creates some employment in disadvantaged rural areas, reduces the cost of the service and meets the villagers' expectations of getting jobs with the project.

Four companies are at present operating on a fee-for-service model in the four concessions. Regulatory, institutional and contractual arrangements for off-grid energy services have been worked out as the part of the programme. Among the achievements is the publication of a service standard for non-grid electricity customers. The standard outlines the service activities and the minimum standards for measuring the quality of service provided by the non-grid service providers. The standards give the National Regulator a basis for evaluating quality of service to non-grid customers.

So far the rollout has often been delayed by institutional and contractual challenges between government and service providers and it is unlikely that the target will be achieved within the next years if the capital subsidies are not paid and installation rates are not increased.

In what Ways did they Claim to Speak for Particular 'Publics'?

Apart from the government the two major publics are the service providers and the rural poor. As part of the RDP the government wanted to fulfill its obligation to provide basic services to the historically disadvantaged population particularly poor rural people. At the same time it had to set framework conditions and guarantee subsidies to attract private business to participate and take up the concessions. The service providers speak for their company, their employees and shareholders.

Table 2: Actors and Expectations Involved in the Solar Home System Project

Actor	Expectation	Speaking for 'Publics'
Energy ministry	Implementing 'electricity for all'	People without access to energy services, poor rural communities, redressing the injustices of the past
Municipalities as electricity distributors	Provide access to limited electricity through solar home systems	Communities without access to electricity
Regulator		Quality of installations Approval of systems Protecting customers
SABS	Setting standards for solar home systems	Mark of approval for SHS manufacturers and installers

Contd...

Contd...		
Eskom	Communicate electrification plans to SHS providers Gain experience with off-grid electricity roll out	Electrification planning Integrate off-grid electricity generation into the system
Service providers	Provide affordable electricity in remote rural areas and grow their business Create a business model for rural electrification and prove its viability, Innovate some aspects such as electricity metering for SHS	Business model development for SHS in rural areas Technology development Employees' and shareholders' interest
Customers	Accessing off-grid electricity	Interacting with the service providers to adapt the system to their circumstances
Villagers and village chief	Finding employment with the project Technical and business training Lighting increases security in area	Improve infrastructure Create employment and training Increase security
Equipment companies	Develop new competencies Create new equipment components Gaining a market share	Employees and shareholders

What were their Expectations of the Renewable Energy Initiative?

The major expectations of the government were to speed up universal access to electricity. The programme targets 300 000 households for SHS, 50 000 for each of the initial six concession areas. Since the SHS only provides electricity for lighting, B/W television and radio it was expected that the service providers would also provide fuels for thermal use such as gas and kerosene. Such fuels are often not available in remote rural areas.

Providing affordable, safe and clean energy to the rural poor is one of the difficult issues of rural development. Service providers are expected to prove that their business models are viable and can energise the countryside. (If the model is economically viable and socially acceptable it can be applied to the billion people all around the world who have no access to electricity.)

3.3 Step Three: Understanding 'Participatory' Decision-making: Negotiating Expectations

How, When and on What Basis were the Different Expectations Negotiated?

In 1998, Eskom and Shell Renewable South Africa announced a joint venture with the objective to provide 50,000 households with SHS in the next five years. This project was widely publicised and politicised and might have influenced the DME to speed up its off-grid electrification programme (Afrane-Okese 2004). In the beginning of 1999 the DME consulted with potential stakeholders, chose the concession model, and advertised the call for proposals. The wide publicity generated by the Eskom Shell Joint Venture created interest in the PV industry and 28 proposals were received. Out of these six were selected and added to the Eskom-Shell Joint Venture.

Part of the programme was to build capacity and to work out the institutional, legal, contractual and regulatory arrangements for the off-grid energy services. This is one of the reasons why the initial phase was very slow.

What (Mix of) Mechanisms (Formal and Informal) were Used? (Systems of Interaction)

There were extensive negotiations between government and service providers. The capital subsidy became a problem when government limited the number of systems it subsidised. It is expected that negotiations to renew these payments to service providers will be announced in the near future. The fact that the service providers continued to operate in the last two years without receiving any capital subsidies indicates that their business model is quite robust. Not all local governments paid the monthly service subsidy of free basic electricity and some paid it intermittently.

Table 3: Forms of Participation in the Solar Home System Concessions

Type	Organizers	Involvement	Purpose
Policy development and planning	Energy ministry	Experts and stakeholders	Create framework conditions for solar electricity roll out
Selecting service providers for the SHS areas	Energy ministry	Experts	Select companies which roll out solar home systems
Further policy and strategy development	Energy ministry	Experts and stakeholders, Regulator, service providers	To develop strategies to roll out and finance solar home systems

Contd...

Contd...			
Community meetings	Service providers	Customers, local representatives	Communicate the project, the technology and the service contract to customers
Informal communication	Customers	Service providers	Complain about the system not working to expectations, ask for clarification how the system and the service contract work.

How were the Interests of Various Actors Aligned?

The interests of the various actors (government, service providers, customers) were diverse, and ultimately only some of the interests were aligned. For example, there was a general interest specifically expressed by government that service providers sell other energy products in their energy shops, while only some do so.

The companies faced major development problems such as poor roads, no transport, no or poor communication. Providing such services is government's responsibility and in this case the lack of basic services in the concession areas contributed to the cost to the service providers. Some houses are inaccessible by car and the installers had to carry the equipment into the valleys (Afrane-Okese 2003). The absence of basic services affects the rural people because it makes income generation and running small business very difficult. There is no access to markets and people find it hard to generate income to pay for their electricity service. These are problems of context and development affecting the project although they are not related to the acceptance of the technology. In effect the technology is acceptable because the areas have limited basic infrastructure. The SHS will provide light, and television will connect the households to the wider world.

What Issues Arose from these Processes?

When the concession areas were awarded, the service providers thought that the basis for allocating the concessions was the fact that electrification was not to reach the area in the near future, but some SHS clients were later connected to the grid. There appears to be a lack of transparent electricity planning and communication. When clients are expecting grid electricity they are generally not willing to accept SHS.

Some customers complained that SHS are only given to poor people, the perception being that PV systems are an inferior technology for the poor. This negative image and the limited power supply were two of the reasons why only 57% of the surveyed households would recommend SHS to others, while 96% recommended grid electricity.

The withdrawal of the capital subsidy is a major issue threatening the viability of the business plan and questions government's commitment to this RE model whereby the service providers were pioneering a new business model embedding the technology into the local economic, social and institutional structure. Some had invested heavily in the new venture.

Some impoverished rural municipalities had other more urgent expenditures and were not able to pay the service subsidy and some paid it irregularly, leaving customers stranded. Customers either had to pay the full service fee or, if they were able to do so, service providers repossessed their systems. This uncertainty of service subsidy also affects the business plan of the service provider.

20,000 clients paid the monthly service fee for the last six or seven years and the percentage of defaulters appears to be no higher than in other similar programmes, indicating an acceptance of the technology and the way it is provided. In some areas, households did not continue with payments, and in such cases the service providers repossessed a few of the installed systems.

As said above, some customers were unhappy with the limited energy and the 'poor' image of the technology. Some did not fully understand the limitations of the SHS as well as their obligations of the service contracts.

3.4 Step Four: From Visions to Actualities

How was the Vision Translated into Action?

In the Eskom-Shell concession the first phase of the project was quickly implemented because 'promises needed to be fulfilled and many pressures towards service delivery to the deprived people existed' (Afrane-Okese 2003). From February 1999 to March 2000 about 6,000 SHS had been installed. The company ran into many problems, Shell and Eskom pulled out and the company was liquidated. Three smaller companies have taken over the concession area and have been providing the services for the last few years. This indicates that the business model is viable, provided the necessary adaptations to accommodate local conditions are made.

Accurate installation figures are difficult to get. It is estimated that 20,000 to 30,000 SHS had been installed under the concession programme by 2004. Assuming an average household size of 4.5, this would imply that about 90,000 people have benefited so far.

Did this Result in Adapting the Initial Objectives of the Vision?

Some of the objectives of the initial vision were achieved. Many had to be altered or completely changed. It appears that the service providers evolved and adapted the details of their original business models successfully, because they have stayed on and are still in business in spite of the fact that government has not paid capital subsidies for two years.

The objectives of the government vision were also adapted. The very poor were excluded because they could not afford the initial installation fee and the monthly service fee. Government thought that the larger companies would be in a stronger financial position to pioneer the new project but one of the largest companies, the Eskom Shell Joint Venture, pulled out after heavy losses and was replaced by three small companies which are still in business.

The technology was adapted and electricity prepayment meters were specially designed and attached to the photovoltaic systems.

How did this Occur Over Time?

The rural electrification project provided 20,000 to 30,000 households with electricity from SHS. These households would not have had electricity without the project. The initial target to roll out 3,00,000 SHS was not achieved in the planned timeframe but if government renews the payment of the capital subsidy this target may still be met in the future.

The Eskom-Shell Joint Venture was the first company to install the SHS and many lessons were learnt in this phase and some of the agreements between the various actors were adjusted. The absence of basic services in the remote rural areas was a major factor increasing the cost of the installation and maintenance. The hope that the utility model would attract larger companies has not been fulfilled.

The expectations of the government that electrification planning, funding and subsidisation would be rationalised have not been fulfilled so far. In some areas grid electricity arrived unexpectedly after SHS were installed and some of the systems were relocated. Other fuels such as gas and kerosene were not offered in all concessions, although this may occur in the future because it seems to be good business.

Government so far only subsidised 20,000 to 30,000 systems instead of the 3,00,000 originally envisaged. The service providers consequently installed fewer systems than they had originally planned – the larger companies 6,000 to 8,000 and the smaller companies probably much less and it is doubtful if any economies of scale can apply.

Some customers wanted larger systems and some of the service providers adopted a flexible approach and provided them.

Generally the service providers have ensured continued services to their customers for five to seven years. The companies generally succeeded in establishing financial and maintenance controls.

4. Key Lessons of the Transition Process

In developed countries RE technologies are most often introduced for environmental reasons to reduce GHG emissions. In this case study the major concern is access to electricity for the poor in remote rural areas and not the environment.

Although the SHS technology is easy to use, the introduction of PV technology in remote rural areas has often been compared to providing space age technology to the least developed populations. In many cases the technology gap and the problems related to service delivery had not been identified as one of the potential major barriers to successful implementation and social acceptance. This knowledge gap extends into two directions. The service provider does not understand the needs and conditions of the customers and the customers do not understand the technology and the often complicated agreements that go with it. The methods for supplying the technology, negotiating government subsidies, etc., are not simple and have led to widespread uncertainty. The provision of SHS has to be backed up by information and training, customer responsive service and maintenance and long-term contractual subsidy agreements with government.

SHS owners are happy having electricity for lighting and media but they still have to use other sources such as fuelwood, kerosene or gas for their greatest energy need, cooking. The monthly SHS service fee has been R58 per household for electric lighting and media only – a high cost for very poor households. The poorest of the poor can afford neither the initial installation fee nor the monthly service fee. In line with its policy of free basic services for the poor, government subsequently proposed a further monthly subsidy of R40/month for SHS users, reducing their monthly payments to R18/month. This makes SHS electricity more affordable to a wider range of poor rural households; but it is difficult to implement this subsidy, because it has to be administered at another government level, local government (in this case, some of which are poor rural district municipalities). Local government leaders may not endorse SHS subsidies if they have higher priority spending needs in their areas. As a result, the R40/month SHS operational subsidy proposed by national government has only reached a few of the concession areas. In one area, this subsidy was started, then stopped, causing quite serious problems for customers and the service provider. Customers residing in different municipalities find it hard to understand why their neighbour receives a monthly subsidy while they do not get any.

In all cases, the installation of SHS has been highly subsidised by the government (R3500 or more per household) and the subsidy may be better used extending the grid. The individual and collective benefits of grid electricity supply are greater than the benefits of SHS services. Nonetheless, SHS have their niche in very remote rural areas, which cannot be reached by grid electricity in the medium to distant future.

The project did not facilitate income generation. Productive end uses for PV systems are known in other parts of South Africa. The addition would have enhanced social acceptance and affordability.

The programme was effective in delivering electricity to the rural people, despite the poorest being excluded. However, considering that the technology, delivery mode, financial and institutional arrangements have been new and in many cases untested, all stakeholders have learned during the process and it is hoped that the next phase of implementation will be easier. It remains urgent to provide energy services to the poor, but PV systems are only suitable in very remote rural areas where the grid will not reach in the future.

The reaction to SHS and the mode of delivery has been ambivalent. The collective benefits include greater security at night because houses and shops are lit. The individual customers are pleased with the limited applications of SHS and enjoyed having lights, watching TV and listening to the radio. They are disappointed that they cannot cook and use heavy electric machinery and consider this a drawback as compared to grid electricity. They still have to pay more for other fuels such as wood, kerosene and gas for their thermal needs such as cooking. Many also do not fully understand the feefor-service model and are often ignorant of the government capital subsidy. In the Eastern Cape study only 57% of SHS-users would recommend a SHS to others while 96% of grid-connected households would recommend grid electricity to others (ERC 2004). SHS were rolled out in remote rural areas, which are generally poor. Some customers felt that the solar systems which have a much more limited range of applications than grid electricity were an inferior technology given only to the poor. This perception created a negative image of the technology.

The service providers showed that their business model to provide photovoltaic electricity to the rural poor was adaptable to the local conditions. They embedded it by and large successfully into the local environment. It appears to be a feasible model for rural electrification, which could be applied elsewhere.

(Gisela Prasad is a geologist by training but has an extensive research and development experience in development studies and has managed several projects including Demonstration of Solar Villages in Lesotho, Rural Technical and Business Community Development Centres, Innovative Rural Action Learning Areas, and Holistic Resource Management. The author can be reached at gisela.prasad@uct.ac.za).

References

Afrane-Okese, Y. and Thom, C. 2001. Understanding the South African off-grid electrification programme, in: ISES 2001 Solar World Congress, 25-30 November 2001, Adelaide, Australia.

Afrane-Okese, Y. 2003. Operational Challenges of large scale off-grid programme in South Africa. NER Quarterly Journal 2003(3), pages 33-52.

Agama 2003. Employment potential of renewable energy in South Africa. Agama Energy, Cape Town.

DME 1998 (Department of Minerals and Energy). White Paper on the Energy Policy of the Republic of South Africa. Department of Minerals and Energy, Pretoria.

DME 2002. Capacity building in Energy Efficiency and Renewable Energy: Baseline Study – Solar Energy in South Africa. Report No. – 2.3.4-13.

DME 2003. White paper on the Renewable Energy Policy of the Republic of South Africa. Department of Minerals and Energy, Pretoria.

DME 2004. Capacity building in Energy Efficiency and Renewable Energy. Report no. 2. 3. 4. – 19. Economic and Financial Calculations and Modelling for the Renewable Energy Strategy Formulation. Department of Minerals and Energy, Pretoria.

DME 2005. Capacity building in Energy Efficiency and Renewable Energy: Renewable Energy Monitoring of Targets. Report No. – 2. 3. 4 – D. Department of Minerals and Energy, Pretoria.

ERC, 2004. Solar electrification by the concession approach in the rural Eastern Cape (South Africa): Phase 1. Baseline Survey. Energy Research Centre, University of Cape Town.

Kotze, IA. 1997. Renewable energy activities in South Africa – PV and Rural electrification. In: Proceedings of the Third OAU/STRC Inter-African Symposium on new, renewable and Solar Energies, 22-24 October 1997, Pretoria, South Africa, Department of Minerals and Energy, pp 10-16, Pretoria.

Kotze, IA. 1998. Photovoltaics and rural electrification in South Africa – Problems and prospects. In: Proceedings of the ISES Utility Initiative for Africa: Initial Implementation Conference, 26-27 March 1998, Midrand, South Africa.

Kotze, IA. 2000. The South African national electrification programme: Past lessons and future prospects. In: Proceedings if the African utility Project: Seminar on Rural Electrification in Africa (SEREA), April 2000, Midrand, South Africa.

Mlambo-Ngcuka, P. 2004. Budget vote speech by the Minister of Minerals and Energy, June 2004, Parliament, Cape Town.

13

Power for Africa: Solar Hybrid Systems for Rural Electrification

This article explains how rural electrification with reasonable renewable energy becomes particularly important in developing and underdeveloped countries as it enables access to clean water, healthcare services, education and economic development.

Developing and third-world countries are the losers of accelerated climate change. These countries that largely rely on agriculture are severely affected by extreme weather conditions and changes in climate conditions. However, they do not have the money to adapt accordingly. And the increasing scarceness and rising prices of fossil fuel oil hamper their development. It is against this background that rural electrification with reasonable renewable energy becomes of particular importance, because decentrally generated electricity allows access to clean water, to health care services, education and economic development.

The Cologne-based company Energiebau Solarstromsysteme GmbH and the development organisation Internationale Weiterbildung und Entwicklung gGmbH (InWEnt, Bonn) developed a concept for electricity supply with photovoltaic systems and vegetable oil generators and successfully implemented

Source: www.solarserver.de © Heindl Server GmbH, Tübingen, Germany. Reprinted with permission.

Solar power system on the roof of a church in Mbinga (Tanzania); Solar modules, inverters and batteries ensure power supply far away from the electricity network.

Photos: Energiebau Solarstromsysteme GmbH.

this hybrid technology in numerous reference projects. In January 2007, they received the "Roy Family Award 2007" for their achievements. This prize is awarded every two years by the Harvard University (John F. Kennedy School of Government). The award recognises successful public-private partnerships in the field of environmental protection. As solar system of the month of February 2007, the Solarserver presents the solar power system combined with a vegetable oil generator operated by the Vincentian Sisters in Mbinga (Tanzania). This solar hybrid system provides clean electricity day and night to 140 people in various institutions run by these sisters.

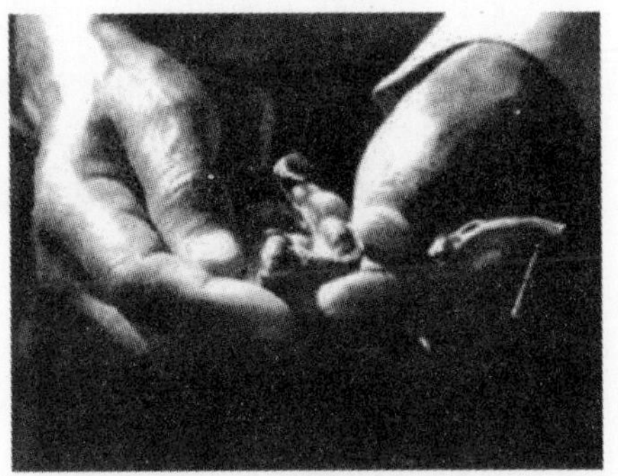

Only a few rural regions in Africa and the Third World are connected to the electricity network. Decentralised power supply from solar and biological energy is an economically and ecologically feasible alternative. Picture, right: nuts of the oil-yielding jatropha plant.

Photos: Energiebau Solarstromsysteme GmbH.

Many Regions without Electricity; Diesel for Generators is Expensive and Uneconomical

Around the world 1.5 billion people must live without electricity. In the rural areas of Africa electricity, if generated at all, is produced with diesel generators, since the establishment of a central power supply with a network of power lines is economically not feasible and far too expensive. Self-sufficient solar power systems that do not depend on a grid offer a reliable alternative. If, however, besides solar radiation diesel fuel is continued to be used as a second combustible, the people will remain dependent on this fossil fuel that is generally transported over long distances. The price for diesel in Tanzania amounts to approx. one euro per litre, i.e. a price that is comparable to that in Europe. Income, however, is significantly lower. The dramatically increased crude oil and diesel prices are an enormous burden for many countries in Africa and Asia, particularly in the rural areas.

People living in developing countries generally pay similar prices for imported crude oil products as Europeans do, but earn significantly less. Picture, right: press to manufacture vegetable oil from jatropha nuts.
Photos: Energiebau Solarstromsysteme GmbH.

Jatropha Plantations Replace Oil Imports

The above reasons led Energiebau Solarstromsysteme GmbH to develop a technology in which generators can also be run on pure vegetable oil that is won on site with a simple mechanical press from the oil-producing jatropha plant.

This plant originating from South America even grows in very dry regions in barren soil. Jatropha nuts contain high degrees of oil but are inedible for humans and animals. Thus, neither on agricultural fields nor among fruits, competition is created between the cultivation of food and these energy-yielding plants. Jatropha oil not only solves the problem of high fuel costs; also the inefficient transport of diesel fuels over long distances as well as the environmental hazards posed by diesel, can be avoided. Furthermore, the rural population can generate additional income from planting and selling jatropha. Thus, on the basis of renewable energy, reliable and sustainable energy supply can be established.

Electricity from Sunlight and Vegetable Oil

The solar hybrid system of the sisters consists of a photovoltaic system for the direct conversion of solar energy to electricity in a generator that runs on pure vegetable oil, as well as inverters and batteries. Since the end of August 2006 the solar vegetable oil system provides electricity to the educational centre of the Vincentian Sisters in Mbinga that consists of twelve buildings. In their convent the "Good Vincentian Sisters" thus lead the way for sustainable development, because they also do without diesel and produce the oil required for the generator themselves from jatropha nuts.

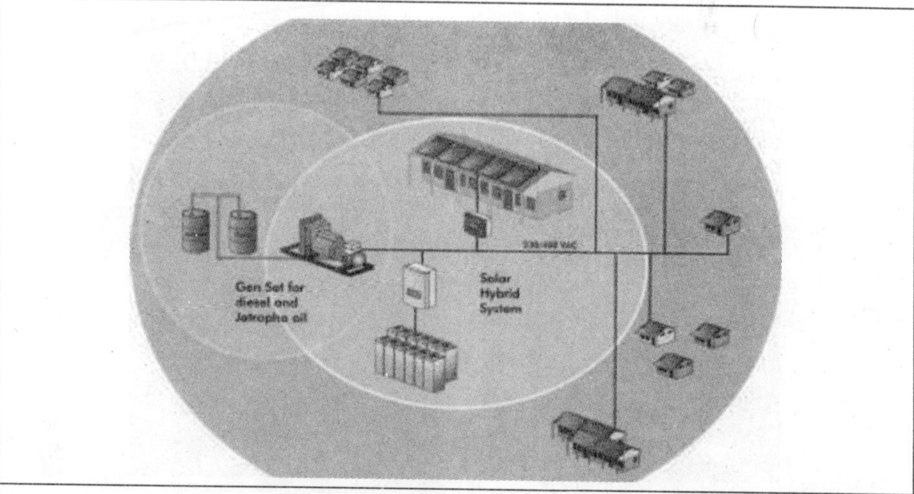

Schematic presentation of a solar hybrid system for basic electricity supply with a (vegetable oil) generator for peak loads.
Graphics: Energiebau Solarstromsysteme GmbH.

If the field of solar modules and the capacity of the battery system are designed accordingly, maximum supply security can be achieved even in times of less sunshine, which however occurs less frequently in Africa than in Central Europe. 81 'Schott ASE 100' solar modules with a total peak performance of 8.1 kW (kWpeak) supply solar direct current to 6 'SMA Technologie AG' inverters. These three 'SMA SI 4248' and three 'Sunnyboy SB 3800' convert the solar power into alternating current, as we know it from our sockets here at home too. The modules cover a surface of approx. 80 m^2 and were mounted on the roof with a 'LORENZ PLUS' supporting structure manufactured by Energiebau Solarstromsysteme GmbH. According to calculations by Energiebau the investment in solar energy production under Africa's sun will have paid off after 10-15 years, so-called energetic amortisation is achieved even faster: within two to five years, the solar power system will have produced as much energy as was required for its manufacturing.

A Model for the Whole of Africa

During assembly of the solar power system, the sisters in Mbinga gained some hands-on experience and acquired knowledge and skills to conduct maintenance work on the existing and future smaller photovoltaic systems in other institutions. They don't want their pupils to lag behind modern technology and unrestricted power supply provides new perspectives in achieving this goal. For the Vincentian Sisters, who are concerned about the economic and ecological development of their country, reliable, economic and ecological power supply is a big gain. "Generating electricity from solar energy and jatropha will be a great help for Tanzania. This can become a model for the whole of Africa," Sister Kaya says.

Sisters assembling solar modules on the church roof; a practical introduction to solar technology. Photos: Energiebau Solarstromsysteme GmbH.

The vegetable oil generator by the manufacturer Kuboto that is only switched on during particularly high power consumption peaks, was adapted by EnergiebauSolarstromsysteme GmbH to run on jatropha oil rather than diesel. Its output is 30 kilowatt (kW) at 1,500 rotations per minute and a consumption of 0.34 litres to produce one kilowatt hour of electricity. The photovoltaics system alone produces a daily output of 35-40 kWh of clean electricity. The vegetable oil generator thus only runs for about 2.5 hours a day. The total solar hybrid system cost the sisters approx. 100,000 euro. Every litre of jatropha oil saves almost one euro in comparison to diesel.

Solar Power Covers Base Load, Generator Covers Peak Loads

During times of low electricity consumption and high solar radiation, the batteries are loaded. Thus the solar power system was not designed to cover peak loads; it serves to cover the basic electricity demand. Peakloads, for example when machines are operated, are covered by the generator.

Generator control panel (left foreground). About two dozen batteries guarantee power supply around the clock.
Photos: Energiebau Solarstromsysteme GmbH.

By combining a solar system with a photovoltaic system, the photovoltaic system can be significantly smaller. The costs of the overall system are thus lower than in a pure solar power system. And the fuel costs are reduced to a minimum.

On the Way to Total Independence from Crude Oil

The list of institutions established in Tanzania by the Vincentian Sisters in the past 40 years is long: it includes a maternity ward, nursery schools, general schools, as well as schools of home economics and agriculture. 18 of these institutions were successfully built up in Africa by sisters from the Swabian village Untermarchtal since 1960. Today seven German and 183 local sisters live and work here. The convent of sisters in Mbinga is fully self-sufficient. The sisters run numerous workshops, e.g. a carpentry workshop, a metal workshop and a book-binding workshop as well as a textile processing workshop in which local sisters receive thorough training in the various trades.

Workshops and schools cannot be operated successfully without reliable power supply. The Vincentian Sisters, however, were no longer prepared to spend an ever increasing share of their financial menas for diesel fuel in order to be able to run generators, particularly since this money, to the largest degree, consists of donations. Eventually high fuel costs even led to certain machines and urgently required equipment only being run on an hourly basis. Now the energy required comes from heaven – and from the earth.

The future with renewable energy: nursery school in Mbinga, jatropha plantation.
Photos: Energiebau Solarstromsysteme GmbH.

Two years ago already the Vincentian Sisters started planting jatropha. With the cultivation of the nut and the production of oil the sisters initiated a local value-adding chain with additional income opportunities for the people in and around Mbinga. In their agricultural school they now teach their pupils about planting and cultivating this energy plant and thereby make a contribution to the sustainable development of the country. Up until today the sisters have planted about 30,000 bushes. A further 30,000 bushes are to follow. "50,000 jatropha plants will allow us to be fully independent from diesel fuel throughout the year and we can then independently produce power for all institutions here in our convent in Mbinga," Sister Kaya explains.

Energy from Renewable Sources: Technically Possible and Economically Feasible

"The project in Mbinga shows that the use of renewable energy in African rural areas that are far removed from the electricity network is technically possible and economically feasible. In the long run the regenerative energy supply systems will drive ahead sustainable and decentralised energy supply in Africa," Michael Schäfer, Managing Director of Energiebau Solarstromsysteme GmbH, explains. In order to achieve this goal, Energiebau and InWEnt organises training seminars on the cultivation of jatropha in numerous regions of Africa.

Solar power system (close-up), Dusk in Mbinga.
Photos: Energiebau Solarstromsysteme GmbH.

Other Reference Projects in Ghana, Tanzania and Indonesia

Specialists from Energie Solarsysteme have already successfully put their concept for decentralised electrification into practice in Ghana, Mali and Tanzania. Furthermore, in co-operation with InWEnt and within the framework of development co-operation a network of partners was established in these countries that can install and maintain such systems. The Federal Ministry for Economics and Technology (BMWi) supported the project in Tanzania through the Deutsche Energie-Agentur (dena) on the grounds of its referential character.

In Ghana (Busunu), a solar hybrid system with a vegetable oil generator supplies a village of 360 houses with electricity. In Matemanga (Tanzania), a water pump is driven by a jatropha generator and a solar power system supplies the local clinic with electricity. In Indonesia, a solar hybrid system with a vegetable oil generator transformed a training centre on Sumba Island into an independent electricity producer.

14

Creating a Credit Market for Solar Thermal: The PROSOL Project in Tunisia

Emanuela Menichetti and Myriem Touhami

This UNEP project explains measures taken to create a level playing field where solar thermal can better compete with conventional energy sources in Tunisia. It also refers to involvement of banks as a successful strategy, since they leveraged financial resources to create a market for solar thermal.

Introduction and Background

The United Nations Environment Programme (UNEP) has been working with the finance sector since the late 1990s to develop innovative mechanisms for the promotion of sustainable energy technologies in developing countries. Through its Renewable Energy and Finance Unit, UNEP has implemented a variety of "financial catalysts" with the aim to lower risks, buy down transaction costs, build capacity and address soft market barriers that constrain sustainable energy technologies growth.

Source: UNEP (2007): Menichetti, E and Touhami, M "Creating a Credit Market for Solar Thermal: The PROSOL Project in Tunisia". Paper contained in the Proceedings of the Third European Solar Thermal Energy Conference (ESTEC), 19-20 June 2007, Freiburg. Reprinted with permission from the UNEP (The United Nations Environment Programme).

The MEDREP Finance initiative represents one of the most recent and successful examples. It has been developed within a manifold programme launched by the Italian Ministry for the Environment, Land and Sea at the World Summit on Sustainable Development of Johannesburg in 2002. The aim of the Mediterranean Renewable Energy Programme (MEDREP) is to expand the share of renewable energy technologies in the Southern Mediterranean region, in order to reduce poverty, combat climate change and achieve long-term sustainability objectives.

A series of projects are being developed in Tunisia, Morocco and Egypt. Below, the structure of the PROSOL project in Tunisia is reported, as well as the results achieved. First, an overview of country-specific conditions is provided.

Framework Conditions and Solar Thermal Potential in Tunisia

Water Heater Typology	% of Stock
Electric	10.40%
Natural gas	8.00%
LPG	78.40%
Solar thermal	3.20%

Tunisia has a significant solar potential, with very high irradiation rates. According to the GIS-based data made available by the European Commission's DG Joint Research Centre (JRC, 2007), the country benefits from 1,700 to 2,200 kWh/m^2 per year. The National Agency for Energy Conservation (ANME) estimates that solar thermal panels could satisfy approximately 70-80% of sanitary hot water needs in the residential sector. So far, solar water heaters cover only 3% of the market in the domestic sector. As one can see, the market is dominated by LPG-fired boilers, which constitute 78% of the existing stock (Missaoui and Amous, 2003).

While sun is an abundant source in Tunisia, the country has scant fossil fuel reserves and its net energy balance has been showing negative values since 2001. In particular, LPG is entirely imported.

According to the data provided by the International Energy Agency (IEA, 2007) in 2004 LPG imports reached 364 ktoe (+22% over 2000 and +164% over 1990). Hot water demand is over 30 million m^3 per year, and is projected to increase up to 70 millions m^3 by 2010. This would imply a further growth in LPG imports, if current energy consumption patterns remain the same. This would translate into a higher deficit in the balance of payments, and in an increase of government's expenditures to subsidize the product. Currently, LPG is subsidized in a measure corresponding to 50% of its real price.

Solar thermal has been repeatedly proposed as a solution to lower the country dependency from imported fossil fuel sources. The first solar thermal energy strategy was developed by the Tunisian government in the 1980s. But only in the period 1997-2001 a real market and technology infrastructure have been developed, thanks to a project financed by the Global Environment Facility (GEF) and the Belgian Cooperation. The support mechanism was based on a 35% capital cost subsidy. At the end of the period, 50,000 m^2 of new solar thermal panels were installed, 8 suppliers (among which 3 manufacturers) and over 130 installers were operating in the market, for a total of 260 new jobs created. Despite these important results, as soon as project funds expired solar water heater sales dropped again.

The PROSOL Project

PROSOL (Programme Solaire) is a 2-year project developed within the MEDREP umbrella. It has a total budget of Euro 1.7 millions, donated by the Italian Ministry for the Environment, Land and Sea.

The project was initiated in 2005 by the Tunisian Minister for Industry, Energy and Small and Medium Enterprises and the National Agency for Energy Conservation (ANME), with the support of the UNEP-MEDREP Finance Initiative.

The objective of PROSOL was to revitalize the declining Tunisian solar water heater market. The innovative component of PROSOL lies in its ability to actively involve the finance sector, and turn it into a key actor for the promotion of clean energy and sustainable development. By identifying new lending opportunities, banks have started building dedicated loan portfolios, thus helping to shift from a cash-based to a credit-based market.

The main features of the PROSOL financing scheme are:

- A loan mechanism for domestic customers to purchase solar water heaters
- A capital cost subsidy provided by the Tunisian government, up to 100 dinars (57 euros) per m^2
- Discounted interest rates on the loans, progressively phased out.

A series of accompanying measures have been developed, which include an awareness raising campaign, a capacity building programme and carbon finance.

Besides UNEP and ANME, key partners include:

- The Société Tunisienne de Banque (STB)
- Two commercial banks (UBCI and Amen bank)
- The State electricity utility STEG (Société Tunisienne d'Electricité et du Gaz)
- Manufacturers, importers and installers of solar water heaters
- Local consultants.

Functioning of the Financing Mechanism

In the PROSOL scheme, loans for solar water heaters are effectively driven by suppliers, who act as indirect lenders of money for their customers. The process begins when a customer decides to purchase a solar water heater from an eligible supplier. It is worth highlighting that only suppliers accredited by ANME can operate within PROSOL. To this end, products must meet a series of technical requirements and performance standards, as set in a manual prepared by ANME. The supplier submits a loan application to a participating Tunisian bank that qualifies the customer's ability to repay the loan. Once the bank approves the loan to the supplier, the solar water heater is installed at the customer's home. The customer pays only the administrative costs of the process.

After the installation, the supplier receives:

- The subsidy payment from ANME of 200 dinars (Euro 114) for a 200-litre system or 400 dinars (Euro 228) for a 300-litre unit, and
- A payment from the bank of 750 dinars (Euro 428) for the 200-litre solar water heater, or 950 dinars (Euro 542) for the 300-litre system.

The customer repays the loan on a pro-rata basis over a five-year term, through the electricity bills issued bi-monthly by STEG. In some cases, however, the extra cost for the solar water heater is compensated by reduced electricity consumption, thus lowering the overall amount to be paid.

Within this scheme, banks do not have any direct contact with the customer, who is the final beneficiary of the loan. They deal instead with solar water heater suppliers. This unusual arrangement provides a double security: loans are officially granted to the solar water heater suppliers who are responsible for repayments, and the consumer cannot easily default because the loan debt is recovered through the customer's electricity bill. In the event a customer does not pay the bill (and hence the solar water heater loan), banks can take action against the solar water heater suppliers that were granted the loan. At the same time, STEG suspends the electricity supply to the customer.

Highlights of Main Features

The most distinctive aspects of the PROSOL financing scheme are: the engagement of banks and the active involvement of the State utility.

In PROSOL, banks play a very important role since they provide the necessary funds to develop the market, accounting for the highest percentage of the finance for solar water heaters.

The engagement of STEG in recovering the loan payments through its electricity bills has provided enough guarantees to banks to extend the loan terms and lower the interest rates. In PROSOL, loan duration was five years instead of the usual three-year term. As for interest rates, the commercial lending rate for similar loan products in Tunisia is 14%. Within PROSOL, banks have agreed to a 7% reduction. Through the MEDREP Fund, UNEP has provided a 7% interest buydown for loans disbursed in the first 12 month and 3% for subsequent loans. This means the rate initially charged to customers was 0% and after 12 months 4%.

The transparency of the system is ensured by independent third party evaluation. At the beginning of 2007, PROSOL was audited by KPMG.

Results and Outlook

Launched in April 2005, the PROSOL project has resulted in an immediate success. In less than one year (April-December 2005), sales reached the record figure of 7,400 solar water heating systems, for a total surface installed of 23,000 m^2. At the end of 2006, an additional 11,000 units were sold, corresponding to approximately 34,000 m^2. The capacity added in the year 2006 was higher than the cumulative capacity installed in the entire period 1985-1996, prior to the GEF project. On an annual basis, the 2006 figure doubled the sales in the previous best year – 2001 (under the GEF programme). Overall, in less than two years the solar water heater market has surpassed 57,000 m^2, representing as much as 50% of the cumulative surface installed from 1985 to 2004. The figures for the first trimester of 2007 (after the end of PROSOL) confirm this encouraging trend.

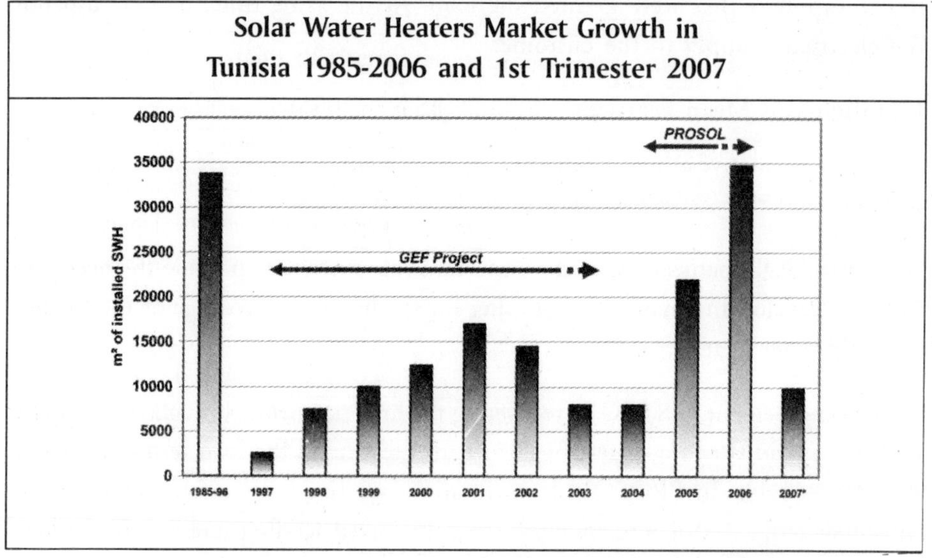

Solar Water Heaters Market Growth in Tunisia 1985-2006 and 1st Trimester 2007

As regards geographical distribution, the majority of solar heating systems are concentrated in the Northern part of the country and in coastal areas.

In terms of equipments, most installations are represented by flat plate solar panels but vacuum tube collectors are slightly gaining market share.

As for financial data, banks have granted loans for more than 6 million dinars (Euro 3.4 million) in 2005 and 9 million dinars (Euro 5.3 million) in 2006.

The number of solar water heater suppliers eligible within PROSOL has rapidly increased, passing from 5 to 9 in few months and stimulating other applications. Today, the number of companies selling solar water heaters is 14, of which 4 are producers. A new manufacturing plant is being established and new technology providers are entering the market.

As far as installers are concerned, their number has reached 384 units, i.e. three times the figure achieved at the end of the GEF project.

These remarkable results have led the Tunisian government to set very ambitious targets. The 11th quinquennial plan has fixed the objective to install 540,000 m^2 in the period 2007-2011, i.e. over 100,000 m^2 per year on average. If this target is met, the annual market for solar water heaters in Tunisia would become comparable to current levels of countries like Spain or Italy, whose population is 4-6 times higher. Solar thermal capacity in operation would reach 48 kWth per 1000 inhabitants, which is comparable to current levels of countries like Germany, Denmark or Switzerland (ESTIF, 2006).

Capacity installed and main economic and environmental benefits over the life cycle of the SWH (15 yrs)	**1996-2004 Around the GEF period**	**2005-2006 PROSOL-related**	**2007-2011 New target**
Solar thermal capacity installed (kWth)	57,000	40,000	378,000
Avoided LPG subsidies (TND)	28,219,000	19,665,000	186,300,000
Primary energy savings (toe)	56,000	39,000	373,000
Cumulative CO2 emissions avoided (tonnes)	307,000	214,000	2,025,000

Moreover, PROSOL has led to an important policy change. In fact, the Tunisian government has made solar water heaters eligible for the energy subsidy that previously was provided only on LPG.

In spite of the considerable strengths of the project, there were also some drawbacks. Two issues need to be addressed:

- The level of exposure of solar water heater suppliers, and
- The lack of dedicated tools for maintenance.

As far as the former is concerned, in PROSOL solar water heater suppliers were the sole party dealing with banks and were obliged to accept responsibility for the loans they had taken out on behalf of their customers. Ironically, a successful vendor was measured by his level of indebtedness. This limit has been corrected in the new "PROSOL2", which was launched earlier this year. In "PROSOL2", end users are directly granted the loan from the bank, thus eliminating the burden for suppliers. The new mechanism has been entirely developed by local actors, and neither UNEP nor other international institutions are involved. This represents a very positive outcome, since it demonstrates that a self-sustaining market and policy decision-making process are being built up.

With respect to the second issue, an audit of half the collective solar thermal surface installed during the GEF project led in early 2007 showed that over one third of the sample was not working anymore because of lack of maintenance. Although within PROSOL specific guarantee requirements are asked to vendors, similar problems might occur if no dedicated measures are adopted.

To avoid malfunctioning in the future, a maintenance cost subsidy has been incorporated in the new Solar Water Heating Loan Facility for the Tourism and Service Sector, in order to ensure long term efficiency of the systems installed, and create expertise in the sector. After the signing of the agreement by the Italian Ministry for the Environment, Land and Sea, the National Agency for Energy Conservation and UNEP the new facility is just being launched.

Accompanying Measures

In order to give visibility to the project, inform customers on the advantages of the mechanism and promote the purchase of solar water heaters, a comprehensive communication plan was developed at national level. The following media were used: TV, radio, posters, brochures.

Moreover, a Training Support Facility was established to build capacity amongst financiers and expand their confidence degree in renewable energy technologies, with the ultimate goal to increase the number of sustainable energy loan portfolios.

Finally, carbon finance is another important component of PROSOL. A Project Idea Note (PIN) was prepared by ANME and submitted to the Designated National Authority in April 2006. A Project Design Document (PDD) is currently under development.

Conclusion

The PROSOL project is proving to be a real success story. In only two years, it has helped achieve a series of long-term goals which go far beyond the actual number of solar water heaters installed.

The first important change driven by the project is the setting by law (Law 82/2005 and decree 2234/2005) of a 20% capital cost subsidy on all new solar water heating installations. This provides a further stimulus to the development of the renewable energy technology market.

In order to improve technology level and decrease costs, the decree 4/2006 has exempted solar water heaters from VAT and decreased custom duties.

These measures help create a more level playing field where solar thermal can better compete against conventional energy sources, like natural gas or LPG. The capacity building and the information exchange have also played an important role in stimulating a dynamic attitude of the Tunisian government, which has set very ambitious targets for solar thermal and has established a comprehensive strategy made of policy, financial and fiscal incentives, awareness raising campaigns (including the "Solar month" campaign), monitoring measures, etc. Engaging the banks has proven to be a successful strategy, since they leveraged enough financial resources to stimulate the creation of a market for solar thermal. As other UNEP projects, PROSOL was relatively small-scale, but has triggered rapid expansion of the solar water heater market. This proves that considerable results can be achieved even with a limited budget, if money is channelled in the proper direction and synergies are exploited. With this respect, an extensive stakeholder consultation process has been carried out and collaboration with all partners involved has been tied in.

Acknowledgement

The authors want to particularly acknowledge the Italian Ministry for the Environment, Land and Sea and the National Agency for Energy Conservation of Tunisia.

(Emanuela Menichetti is Associate Programme Officer (presenter) and Myriem Touhami is Programme Officer at Energy Branch, United Nations Environment Programme-Division of Technology, Industry and Economics (UNEP-DTIE). They can be reached at Emanuela.Menichetti@unep.fr, and myriem.touhami@unep.fr respectively.)

References

(1) Missaoui, R. & Amous, S. (2003). Financing the development of the renewable energy in the Mediterranean Region. Baseline study for Tunisia, Tunis.

(2) ESTIF (2006). Solar Thermal Markets in Europe, Brussels.

(3) International Energy Agency – IEA (2007): Energy Balances of Non-OECD countries, OECD/IEA, Paris.

(4) Joint Research Centre – JRC (2007). Solar irradiation data utility (available at: *http://re.jrc.ec.europa.eu/pvgis/apps/radmonth.php?lang=en&map=africa)*

15

Solar Power in India: The Nascent Stage

Mohan Kumar Devarajan

Solar energy is applied to different industrial and community purposes. India is blessed with abundant sunlight. The vital efforts taken during the past two decades for producing solar power are now providing good results and is realized by people in all walks of life. Various projects on utilizing solar energy have been implemented in India with success. This natural power source if harvested on a large scale will be very helpful in alleviating India's energy constraints.

Energy is neither created nor destroyed; it can only be transformed from one form to another.

– Law of Conservation of Energy.

People have been utilizing sunlight for common purposes such as for drying of clothes and growing crops/plants since thousands of years, but recently people have also started utilizing the sunlight for generating power. The importance of solar power is recognized across the various developing nations to meet their energy requirements for the purpose of generating electricity.

This article earlier appeared in the book "Energy Security in India: Current Scenario", published by the Icfai University Press. *© The Icfai University Press. All rights reserved.*

The term solar power means a renewable energy source which is available abundantly in sun rays. There are a number of methods of producing energy from Sunlight. Each sun ray emits 1,366 watts/meter^2 at the distance of the earth's orbit, but less at ground level. The sun is 150 million kilometers away from the earth's surface and is powerful source of radiating energy. Even if the minutest radiation of the sun's light hits the surface of the earth (around a hundredth of a millionth of a percent) it is sufficient to meet all our power needs many times over. In fact, every minute, an adequate amount of energy arrives on earth to meet our energy requirements for a whole year if only it is properly utilized through innovative technologies.

Solar energy has been applied in many traditional building methods for centuries; recently it has gained interest in the developed countries also as the environmental costs and limited supply of other power sources such as fossil fuels are realized. It is already being used in wide areas of other geographical locations such as in remote locations due to lack of other power supplies.

Advantages and Disadvantages of Solar Energy

Advantages

Solar energy is a pollution-free, efficient, and sustainable form of energy. There is a wide gap between the demand and supply of solar photovoltaic (PV) energy, with lead times ranging between 3-6 months from the suppliers. Solar energy becomes a cost-effective alternative, when the cost of supplying electricity to remote locations is very expensive in this modern world.

1. The cost of PV systems has reduced 25 times less than the actual price over the last 20 years. It is the most cost-effective energy source available in remote locations such as cabins and resorts.
2. Solar energy is free—it doesn't need any fuel for generating the power from sunlight.
3. It doesn't produce any waste either in the form of air or solid.
4. It neither pollutes the environment nor causes any harm to the surroundings.
5. In countries where it is sunny, the solar power could be exploited to the maximum when there is no other alternative to generate electricity to meet the ever increasing energy requirement.

6. It is handy for low-power utilities such as solar powered garden lights, battery chargers, water heaters..., etc.
7. Solar power system is less expensive in producing power for even remote villages. In some cases, it can cost over $20,000 to run in public utility power lines.
8. Solar power is one of the friendliest sources to the natural environment, as it generates electricity through means and technologies, without emitting any harmful radiations.

Disadvantages

1. It is not effective during the night time.
2. It is quite expensive to build solar power stations. Solar cells cost a huge price compared to the quantity of electricity they'll produce in their lifetime.
3. It is only suitable for the sunny climate.
4. Though there is significant reduction in the cost of solar energy, it is still higher compared to other electricity sources.

Solar Power Generation

Power from the sun arrives on the surface of the earth as heat and light radiations. These radiations are the outcome of the sun's constant nuclear fusion of hydrogen nucleus. This process of fusion creates helium nucleus along with the emission of a large amount of energy. The emitted energy is called electromagnetic radiations (light is a specific frequency range of this radiation). The temperatures of sun's radiations are measured over 6,100 degrees Celsius. This is cooler when compared with the aura and core of the sun that burns at several million degrees Celsius. Only a tiny fraction of the sun's radiations reaches the surface of the earth. The average amount of energy that reaches the earth's surface in a day is 200 W/m^2.[1] This means that the average home can fulfill its electricity requirement more than enough through the solar radiation available in its own roof space. In fact, each day, more energy accumulates on the Earth's surface from the sun than would be consumed by the global population in 27 years.[2]

---

1 Stefan K Estreicher and Mark Holtz, Texas Tech University, "Solar Energy". *http://jupiter.phys.ttu.edu/corner/1998/jan98.pdf.* Accessed June 19, 2001.

2 BP Solar, "A Technology for Today and Tomorrow". *http://www.bpsolarex.com/3rd-Section.html.* Accessed June 19, 2001.

There are two types of technologies linked with solar power. These technologies are divided into two groups. They are given as follows.

1. The first group uses the sun's radiations to produce heat. It is called solar thermal technology. Solar thermal technology includes solar concentrator power systems, flat plate solar collectors, and passive solar heating.
2. The other solar power technologies directly convert solar radiation into electricity through the photoelectric effect by using photovoltaic cells (also known as PV).

Solar Thermal Technologies

a. *Concentrating Solar Power (CSP) System:* It produces electricity via heat radiated by the sun. Concentrating Solar Power (CSP) technologies accumulate solar energy to produce high temperature heat, which is converted into electricity. The three most advanced CSP technologies are Parabolic Troughs (PT), Central Receivers (CR) and Dish Engines (DE). CSP is suitable for today's modern efficient power plants. CSP can be used as an alternative solar heat for fossil fuels, fully or partially, to reduce emissions and provide additional power at peak times. DE is well suited for distributed power, from 10 kw to 10 mw, while PT and CR are generally used for larger central power plants, whose capacity is in between 30 mw and ≥200 mw.

   The solar resource for producing power from concentrating solar power systems is bountiful. The amount of power generated by a Concentrating Solar Power (CSP) plant relies on the total amount of direct sunlight. Like concentrating photovoltaic concentrators; these technologies use only the direct-beam sunlight, rather than diffuse solar radiation. Concentrating solar system uses mirrors and lenses to muse and focus sunlight onto a receiver that gets accumulated at the system's focal point. Then the receiver absorbs and converts the sunlight into heat. Subsequently, this heat is transported by means of a heated fluid (either water or molten salt) through pipes to a steam generator or engine where it is changed into electricity.

b. *Flat plate solar collectors*: They are usually large horizontal boxes with one or more glass covers. Dark colored metal plates are used inside the boxes for absorbing the heat. Then air or liquid is allowed to pass through the tubes and warmed by the heat stored in the metal plates. This system is extensively used for providing hot water to households.

c. *Passive solar heating design*: This method uses large south-facing windows and building materials that take up the sun's thermal energy. Passive solar method can even be used to cool a building using natural ventilation. The simplest and most common feature of the passive solar technologies is absorption of solar energy directly from the sun. A direct gain system includes south-facing windows and a large mass, usually made of stone, brick, or concrete, placed within the space to collect the sufficient direct sunlight in cold weather and the least direct sunlight in hot weather. The result is that in cold weather the large thermal mass in the room absorbs more solar energy and emits heat throughout the room. During warmer times, the thermal mass absorbs only the warm air already in the room, due to its position away from the windows. This circulates cool air in warmer seasons and hot air in cooler seasons.

Photovoltaics

The second main method for capturing the sun's energy is via the use of photovoltaics. Photovoltaics (PV) exploit the sun's photons or light radiations to create electricity. PV technology works on the principle of photoelectric effect first discovered by the French physicist Edmund Becquerel in 1839.

The photoelectric effect takes place when a beam of UV light, composed of photons (quantized packets of energy), hit one part of a pair of negatively charged metal plates. This causes electrons to be "freed" from the negatively charged plate. These negative charges are then attracted to the other plate by electrostatic forces, which subsequently cause electrical current, due to the flow of electrons. The flow of negative charges can be collected in the form of Direct Current (DC). Then, this DC is converted into Alternating Current (AC), which is the electrical power that is most commonly utilized in all houses/flats for domestic purposes.

Solar Power Projects in India

India is blessed with abundant sunlight, water and biomass. The vital efforts taken during the past two decades for producing solar power are now providing good results as it is realized by the people. Today, public are more aware of the benefits of renewable energy, particularly the need for decentralized energy in the remote villages and in urban or semi-urban centers. India has the world's largest programme for renewable energy. The government of India formed a separate department for renewable energy source, called Department of Non-conventional Energy Sources (DNES) in 1982. In 1992, a full-fledged Ministry of Non-conventional Energy Sources was set up under the governance of the Prime Minister.

Box 1: Core Activities of the Ministry of Non-Renewable Energy Sources

- Promotion of renewable energy technologies.
- Creating an environment favorable to promote renewable energy technologies.
- Creating an atmosphere conducive to their commercialization.
- Appraisal of renewable energy resources.
- Research and development, demonstration.
- Extension.
- Production of biogas units, solar thermal devices, solar photovoltaics, cook stoves, wind energy and small hydropower units.

Source: Renewable energy scenario in India, http://www.indiasolar.com/ren-india.htm

Solar collectors have raised the initial investment and the related capital cost in comparison with the fuel-fired power plants. Interests for extra debt and equity, insurance costs, taxes and custom duties have to be paid, additional land has to be purchased and further the staff have to be employed. In contrast to this, fuels are acquired without any interest or insurance rates, and are often free of custom duties and taxes or even subsidized by the government. Thus, CSP requires a start-up investment to enter the market and subsequently to follow the learning curve. It was accomplished by the implementation of 'Spanish Renewable Energy Act' at the end of 2002. For developing countries like Mexico, Morocco, India and Egypt, a grant by the Global Environmental Facility (GEF) of approximately 50 million Euros per plant is provided to the applicable solar projects.

It is estimated that India receives solar energy equivalent of 20 mega watts per square kilometer of its surface area and is committed to gradually increasing the production and exploitation of new and renewable sources of energy, particularly solar energy. The solar energy is being consumed through both available technologies, such as the solar thermal route and the solar photovoltaic route. It has been estimated that India would produce 130 mws capacity of solar power through CSP technologies by 2012 (refer Appendix 1).

Solar photovoltaic cells are known to be a useful power source for domestic purposes like lighting, pumping of underground water, telecommunications, etc. They have also been utilized as power plants for meeting the entire electricity needs of isolated villages, hospitals or lodges. Based on single crystal solar cells of silicon, numerous types of solar devices were deployed in India. These devices include solar lanterns, domestic lights, street lights, community lighting systems, power for the telecommunications equipment installed on the offshore oil exploration and mining platforms, rural telephone systems and the energy deprived population mostly lives in places, where expanding the national power grids would be very expensive, both in financial and energy terms. The solar photovoltaic power is used as an influential energy source in rural and remote areas. In the Sagar island of West Bengal the 26 kilowatts capacity of solar photovoltaic plant is producing the entire electricity requirements of all 300 homes. Similar initiatives are also implemented successfully in some regions of the southern India.

In 1998, two 100 kilowatts capacities of grid-interactive power plants were established in the province of Uttar Pradesh. Private sector firms were fully involved in these projects. Similarly, a 100 kilowatt power generation system has been installed in the state of Maharashtra and is being operated, by a private Indian company. Several companies in the private sector have also installed confined solar photovoltaic power plants in India to meet their own electricity demands. The significant incentives are being offered by the government of India to the private sector and individual citizens in encouraging their participation in promoting the use of solar energy. It is constantly supported by the government through a separate financing agency exclusively called the Indian Renewable Energy Development Agency.

The solar energy equipment shops, "Aditya", were being established in major cities and towns, in collaboration with several manufacturers' associations and non-governmental organizations. During the 'World Solar Programme', organized on October 15, 1998 Mr. Aslam Sher Khan, M.P said,[3] "Market orientation is being imparted to make renewable energy programmes commercially viable and sustainable, limited budgetary funds being used only for selected demonstration projects."

The Project Initiated in the Himalayas

The Himalayan mountain villages suffer severe winters where the temperature drops down below -40°C. For half the year, almost all families have to spend most of their time in one room, crowded around a wood burning stove, or an open fire, and unfortunately have to share their space with animals for warmth. They also deal with the shortage of water supplies that are usually frozen during the winter season. Kerosene lamps provided dim lighting, but only to those who could afford to buy the fuel and did not mind the cost and the travel of two days on foot. During the snowy winter season life became unbearable to them.

During 1993-2003, the Barefoot college of Rajasthan had played a vital role in introducing solar technology to remote and inaccessible villages in the Himalayas. This project had made the impossible project to a feasible one, by the implementation of solar equipment in the remote villages of Himalayas with the help of appropriate training; the poor and rural communities have benefited. This project had trained illiterate and semi-literate villagers as 'Barefoot Solar Engineers' (BSEs), at its Barefoot College in Rajasthan. After their training, villagers returned to their home villages and installed solar units for their own communities. They had taught their own communities the competent knowledge and skills required for repairing and maintaining this solar service.

In the Himalayas, 15,000 people are now using fixed solar units, or solar lanterns, since then life has changed dramatically. Solar water heaters are used to prevent water freezing in the cold winters. The solar lighting systems available in these villages have made this vision pretty clear even during the dark wintery season. Children go to school during winter, thanks to the construction by the Barefoot Solar Engineers (BSEs) of 'solar passive' houses. These are south facing houses constructed with mud and straw fitted with glass panels. The houses

---

3 Agenda item 158: World Solar Programme 1996-2005, *http://www.un.int/india/ind222.htm*

gather and keep heat during the day and are able to prolong the warmth at night, maintaining a temperature of 20°C even when the temperature outside has dropped below -10°C.

Funding for this Project

During 1997-2003, village energy and environment committees had successfully managed this project, and the villagers had made monthly contributions towards repair and maintenance accumulating Rs.1,490,226.33 during this period. This was contributed voluntarily by the village people who routinely earn less than Rs.45 a day.

For the same solar project the Barefoot College had received funds from:

- The Government of India
- Plan International
- Hivos (Netherlands)
- German Agro Action (Bonn)
- The European Commission
- UNESCO Paris.

This project has been implemented in 6 states along the Himalayas since 1990. A total of 110 Barefoot Engineers, of which 22 are women, had been trained during this time. These trained barefoot engineers are looking after the five thousand solar lighting systems.

The Barefoot college of Rajasthan received the "Ashden Awards"[4] in 2003 for maintaining the health and welfare of the people by the implementation of solar technology in the remote and inaccessible villages in the Himalayas. India is one of the few countries that has initiated research and development, with a view to exploiting new and renewable energy sources, before the energy crisis of 1973. As an outcome of continued efforts, a substantial Research and Development and manufacturing sector has appeared in the country for the design and production

4 Ashden Awards is an annual competition to identify and reward organisations which have carried out truly excellent, practical, yet innovative schemes, demonstrating sustainable energy in action at a local level.

of non-conventional energy supply equipment. India also has one of the world's largest programmes for renewable energy.

NGOs/Private Concerns Initiatives

The Bangalore Salesian Society (NGO), Bangalore SELCO (private company) and Winrock International (funding agency) have formed a basket fund. Winrock provided the initial investment to establish a revolving fund, which was then used to light up to 60 households of scheduled caste and tribal families living on the border of the states of Kerala and Karnataka. These families were earning an insufficient income either by working as farm laborers or by doing other small labor jobs. The villagers earned their income by weaving baskets with reeds which were collected from the forest while returning from their farm works. Unfortunately the families were able to weave only one basket per day, as they could not work at night since the entire village didn't have any power supply. The key challenge before them was extending their working hours. Access to the conventional power supply grid was not feasible due to the remoteness of the area and unavailability of power supply. The Bangalore Salesian Society promoted the availability of solar power in that remote village, with the support of SELCO (private company) and Winrock International.

Box 2: Vital Role

- The NGOs have spread the concept and convinced the villagers of its merits. They also actively participated in managing the fund and collecting the repayments.
- The funding agency is providing the needed capital to set-up the revolving fund.
- The private company (SELCO) acts as system supplier and service provider.
- The beneficiaries are the tribal families who share the risk through repayments, and othermonetary benefits earned from their basket weaving.

This solar project was made successful through a 'revolving fund' mechanism, where each household was given a solar home system, with one condition being that the money earned from every third basket would be paid to the revolving fund. This initiative was established in 1993 with 60 households and later expanded to 120 households by 2003 within a decade. The important feature of this project is the loan given to the NGOs without any interest, but with 3-5 years repayment period, primarily linked with productivity. The NGO received this money as grant, which was then utilized for the establishment of the revolving fund.

Box 3: Success Factors

1. Interest of the NGO in the field of work (RET promotion)
2. Private players sharing the 'risks'
3. Mechanisms for gradually reducing risks and transferring of risks.

Source: "Energizing Partnerships: Experiences from India", Preeti Malhotra, British High Commission, New Delhi, http://www.adb.org/NGOs/annex1003.asp

A similar initiative has been taken by the Indian Canada Environment Facility (ICEF) and by Tata Energy Research Institute (TERI) for the purpose of providing renewable energy technologies to the remote villages. Originally they formed the Energy Service Network (ESN) concepts for the same purpose. The ESN comprised NGOs, dealers, retailers and community groups that were then consolidated into a commercial collaborative to retail, assemble, service and market RET in rural areas. The role of various players in ESN is clearly illustrated as follows:

1. TERI: Project implementing agency—to catalyze the development of the ESN.
2. ICEF: Acts as the Project-funding agency.
3. Project office (private entrepreneur): As an effort by the arm of TERI, the project office has shaped a niche for itself, different from the role of TERI. In this view, it has an entrepreneurial role and operates as a cost/profit center. It takes a vital role in brand building, technology infusion, market research, local assembling, and logistics for the ESN.
4. NGOs assist the project office in all its primary functions. Some of the NGOs in the ESN also play an entrepreneurial role.
5. Dealers/sub dealers are considered as the sales and service arm of the ESN.

Similar kinds of projects have been commenced in Punjab and Lakshadweep too. In Lakshadweep, the BHEL has taken the initiative in constructing the country's largest solar power plant for the uninterrupted power supply; so far the island has been fully relying on diesel power for their electricity requirements (refer Appendix 2).

In Punjab, the Solar Energy Program is implemented for supplementing thermal energy requirements by producing solar energy at various temperatures by directly converting it into heat energy. A total of 80 solar cookers, 310 two-light-and-one-fan system, 250 solar street lights were established by the Punjab Energy Development Agency during 2003-04. The solar power projects are also being encouraged by most of the banking sectors and non-banking financial services for the benefit and welfare of the society.

Some 19 banks and other NBFCs were actively involved in promoting the use of the solar water heating system through a unique scheme called "Accelerated Programme on Solar Water Heating (APSWH)" which was launched by the Ministry of Non-conventional Energy Sources in August 2005, (refer Appendix 3). The Indian Renewable Energy (RE) sector is diversified and provides strong business opportunities to US companies. The market potential in India for RE business is projected at US$500 million and is increasing at an annual rate of 15 percent. The major areas of investment are: solar energy, wind energy, small hydro projects, waste-to-energy, biomass and alternative fuel. The new RE policy of the Government of India (GOI) focused at generating 10,000 mws through renewable and non-conventional source by 2012 is expected to increase further the growth rate of this sector.

There are certain factors responsible for the growth rate of renewable energy sector in India. Indeed the demand for electricity is greater than what is actually supplied. Hence, to meet the actual demand, the renewable energy sector is utilized. It is very influential in raising the use of alternative fuels or energy for the domestic requirements like petrol with Octane, gas cylinders as an alternative for petrol in the cars (particularly it is implemented in all the private and public vehicles in New Delhi to reduce the pollution). The secondary factors supporting the need for the renewable energy sector in India are as follows:

- India possesses enormous RE resources like the solar, wind, biomass materials, urban and industrial wastes and small hydro resources.
- Low gestation periods for establishing RE projects with quick returns.
- Favorable government policies.

- Several financing options available for capital equipment.
- Increasing awareness among the industry people in safeguarding the natural environment, which has made the corporate responsible for preserving nature.

Over 150,000 square meters of solar energy collector area has been established in the country for solar water heating in domestic, industrial and commercial sectors by the creation of cumulative installed collector area over one million square meters. A solar steam cooking system comprising of 20 dishes of 12.6 square meters area each has been commissioned at the Global Hospital and Research Centre, Mount Abu. The system generates around 1,200 kg of steam per day which is being used for the purpose of cooking, sterilization and laundry purpose.

On April 16, 2006, Shri Vilas Muttemwar, Minister for Non-Conventional Energy Sources, announced a statement that his ministry would expend about Rs.2,000 crore to light up around 10,000 remote villages apart from remote rural community via renewable energy under the "Rajiv Gandhi Grameen Vidyuti Karan Yojana" scheme for the purpose of providing electricity to rural households. The electrification of those remote villages would be implemented through the Remote Village Electrification Programme of his ministry during the next three years and that ensured to provide electricity/lighting to over one million rural households.

Conclusion

Most parts of our country have approximately 300 sunny days in a year. The enormous natural resources available are not fully utilized by India, be it solar or wind power, or other biofuels. Some of the private companies, NGOs and other private institutions specialized in solar energy must utilize this opportunity for the benefit and welfare of the society in addition to their profit motive. The World Bank report released in April 2006 revealed that 65% of the population lives in rural India and there lies the enormous market potential for renewable energy technologies.

A few private concerns and NGOs with foreign organization's initiative, have implemented the rural electrification process through solar technology in some remote villages of Karnataka, Kerala and Himalayas. The role of the government policy plays a vital part in the rural electrification process through solar technology in terms of incentives/subsidies offered to the private concerns and NGOs. The government has to encourage the FDI flow into the energy sector through an exclusive strategy, which fits better into our system through efficient foreign policies exclusively for renewable energy technologies, similar to the one which is implemented by the European Union Energy Initiative (EUEI), i.e., by taking a regional stand to explore the need for specific projects and areas for financial co-operation rather than establishing the bilateral relations with other developed countries.

In a three day workshop held on "Access to energy for sustainable development and policies for rural areas" organized by the GFSE Regional Workshop on November 29, 2004 at Paro, Kingdom of Bhutan, where S K Chopra, ex-minister of Non-Conventional Energy Sources, opined that certain drawbacks were faced by Integrated Rural Energy Programme (IREP)[5] in India, such as lack of awareness of technology development, lack of technology demonstration, financial constraints, uncertainties about the cost-effectiveness of energy efficient options, market imperfections and absence of infrastructure and stakeholder communication. He had highlighted that 'rural electrification' is the primary issue, since rural communities are depending on non-commercial fuels and biomass. He suggested that energy policies in India should encourage rural solar power, improved cooking stoves, and active involvement of local institutions, as well as coordination with other rural development programmes.

(Mohan Kumar Devarajan is a Research Associate at Icfai Research Centre, Chennai. He can be reached at mohanvsi@gmail.com)

References

1. "Solar Power is Energy from the Sun", by James Wilson, *http://www.darvill.clara.net/altenerg/solar.htm*

2. "Solar Energy Advantages", *http://www.solarsense.com/Advantages/Advantages.html*

3. "Solar Power", *http://www.crest.org/articles/static/1/995469913_2.html#types*

4. "Concentrating Solar Power", *http://www.rurdev.usda.gov/rbs/farmbill/ConcentratingSolarPower.doc*

5 IREP aims to develop capabilities to prepare for implementation of micro-level rural energy plans and projects, and to provide a delivery system for effective technology transfer and utilization of energy for rural development.

5. "Concentrating Solar Power Now", *http://www.dlr.de/TT/system/publications/CSP-Johannesburg.pdf*
6. "Solar Energy to Meet Basic Needs in Himalayas", *http://www.ashdenawards.org/winners/barefootcollege*
7. "Agenda Item 158: World Solar Programme 1996-2005", *http://www.un.int/india/ind222.htm*
8. "Industry and Energy Sector". *http://punjabgovt.nic.in/economy/industryenergysector.htm*
9. "India Energy Market", *http://www.buyusa.gov/kern/indiaenergyreport.html*
10. "Energizing Partnerships: Experiences from India", Preeti Malhotra, British High Commission, New Delhi, *http://www.adb.org/NGOs/annex1003.asp*
11. GFSE Regional Workshop, "Access to Energy for Sustainable Development and Policies for Rural Areas", Final Summary, Emily Boyd, Ph.D. Fiona Koza *http://www.iisd.ca/download/asc/sd/sdvol93num2e.txt*

APPENDIX 1

CSP Market Areas and Lead Near Term Opportunities		
Sl.No.	**Country**	**MW Capacity by 2010**
1	Algeria	130
2	Australia	100
3	Brazil	100
4	Egypt	100
5	Greece	50
6	India	130
7	Iran	130
8	Israel	200
9	Italy	100
10	Jordan	130
11	Mexico	300
12	Morocco	150
13	Namibia	100
14	South Africa	100
15	Spain	200
16	United States	200
	Total	**2220**

Source: "MarketPull/Technology Push – Where to Invest", http://www.nrel.gov/ncpv_prm/pdfs/33586097.pdf

APPENDIX 2

BHEL Commissions Solar-Diesel Hybrid Power System

Lakshadweep, India (February 27, 2006): Indian solar cell manufacturer, Bharat Heavy Electricals Limited (BHEL) has successfully commissioned the country's largest solar-diesel hybrid power system at Bangaram Island in Lakshadweep.

Being a sought-after tourist destination, Bangaram will benefit immensely with the commissioning of the system by BHEL's Electronics Division, Bangalore, which will efficiently supply power, besides considerably reducing the consumption of high speed diesel and conserving the ecology and environment of the tourist island. Boasting of one of the world's most eco-friendly beaches, with facilities for scuba diving, snorkeling & water sports, Bangaram Island attracts a number of international tourists which calls for reliable, uninterrupted power supply for cottages, restaurants, community lighting, etc.

Until now, Lakshadweep administration has been depending entirely on diesel power plants, which posed a threat to the environment due to air as well as groundwater pollution. Transportation of diesel from the mainland (Kozhikode) also depended largely on sea conditions.

Besides, storage of large quantities of diesel on the tiny island was a major safety hazard. To overcome these problems as also to meet the increasing demand for power, the Electricity Department of Lakshadweep embarked on a mission to use solar photovoltaic technology.

Initially, a 10 kw solar power plant was commissioned by BHEL in the island. Based on its success, Lakshadweep administration decided to commission a 50 kw solar diesel hybrid power plant to ensure round-the-clock power supply.

Source: http://www.solarbuzz.com/News/NewsASPR136.htm

APPENDIX 3

19 Banks and NBFCs Join the Scheme on Accelerating Use of Solar Water Heating Systems – Secrĕtary, MNES Reviews Progress of APSWH

Wednesday, April 26, 2006: The new scheme launched by the Ministry of Non-conventional Energy Sources to accelerate development and deployment of solar water heating systems has registered encouraging response. So far 19 banks and Non-Banking Financial Companies (NBFCs) recognized by Reserve Bank of India have joined the scheme to provide soft loans to the beneficiaries in domestic, institutional and commercial sectors which is quite more than only 7 public sector banks who participated in the earlier scheme. This was revealed at a meeting chaired by Shri V Subramaniam, secretary, Ministry of Non-conventional Energy Sources here today to review the new scheme to implement "Accelerated Programme on Solar Water Heating (APSWH)" which was approved in August, 2005.

The banks who are participating in the scheme include Canara Bank, Bank of Maharashtra, Punjab National Bank, Syndicate Bank, Union Bank of India, Andhra Bank, Vijaya Bank, Dena Bank besides some NBFCs and cooperative banks. The new scheme aims installing solar water heating systems at one million square meters of collector area during its two years, i.e., 2005-06 and 2006-07 of which about 4 lakh square meters have been achieved and another 6 lakh square meters are likely to be achieved during 2006-07. In all 15 lakh square meters of collector area have been achieved in the country so far.

Source: mnes.nic.in/pressrelease.htm

Focus on Solar Energy in India

– Aarti Chirag Bhoorat

On 30 June, 2008, the Prime Minister of India Dr. Manmohan Singh released the National Action Plan on Climate Change. The focus of the action plan was on eight National Missions, the first being "Solar Energy". According to the Prime Minister its success can change the face of India by helping in diversifying our national energy requirements to meet the challenge of climate change.

A lot of initiatives are being taken by the Ministry of New and Renewable Energy (MNRE) in development and promotion of solar energy. Two advanced photovoltaic facilities namely, a large area Sun Simulator and a Photovoltaic Concentrator Module Test Bed were inaugurated at Gwalpahari, Haryana, on 14 May 2008. Many un-electrified villages like Dageriya in Dahod district of Gujarat have been developed into self-managed Solar Photovoltaic power units. A remote tribal village Tobakla in Tripura has been developed with a non-conventional solar power plant under Rajiv Gandhi Grameen Vidhyutikaran Yojana. An integrated solar energy park will be coming up in West Bengal which will have a capacity of 30-MW. The Haryana Electricity Regulatory Commission (HERC) has proposed attractive tariffs for generation of power through solar energy to attract top corporate houses like Reliance Industries Ltd., Epuron Renewable Energy Power, Admire Energy Solutions (Moser Baer Company), etc.

Great business opportunities are being seen in the possibilities of developing green buildings. Initiative for promoting this concept in our country has been taken up by India Green Building Council (which is part of CII-Godrej Green Building Council). This proposition is coming up in the form of residential complexes, hospitals, educational institutions, laboratories, corporate offices, etc. One of the technologies which will be used in such buildings is BIPV (Building Integrated Photovoltaic). West Bengal has formulated tariffs for promoting use of small-capacity BIPV systems. India's first solar housing complex was built under these tariffs by WBREDA (West Bengal Renewable Energy Development Agency) at New Town Kolkata.

Thus, India is speedily gearing up to reduce its per capita GHG (Green House Gas) emissions. Various promotional schemes, tariffs, soft loans and technological developments will surely boost production and application of solar power in the near future.

Sources:

1. Govind Singh (June 2008), "national Action Plan on Climate Change Launched: Solar Energy to Change the Face of India".
2. Akshay Urja, Volume 1, Issue 6, May-June 2008.

© *The Icfai University Press. All rights reserved.*

INDEX